Polymer Research
in Microgravity

ACS SYMPOSIUM SERIES **793**

Polymer Research in Microgravity

Polymerization and Processing

James Patton Downey, Editor
National Aeronautics and Space Administration

John A. Pojman, Editor
The University of Southern Mississippi

American Chemical Society, Washington, DC

Library of Congress Cataloging-in-Publication Data

Polymer research in microgravity : polymerization and processing / James Patton Downey, John A. Pojman, editors.

    p. cm.—(ACS symposium series ; 793)

    Includes bibliographical references and index.

    ISBN 0–8412–3744–1

    1. Polymerization–Effect of reduced gravity on—Congresses.

    I. Downey, James Patton, 1960- II. Pojman, John A. (John Anthony), 1962- III. Series.

QD281.P6 P634   2001
547′.28—dc21   2001022567

The paper used in this publication meets the minimum requirements of American National Standard for Information Sciences—Permanence of Paper for Printed Library Materials, ANSI Z39.48–1984.

Copyright © 2001 American Chemical Society

Distributed by Oxford University Press

All Rights Reserved. Reprographic copying beyond that permitted by Sections 107 or 108 of the U.S. Copyright Act is allowed for internal use only, provided that a per-chapter fee of $20.50 plus $0.75 per page is paid to the Copyright Clearance Center, Inc., 222 Rosewood Drive, Danvers, MA 01923, USA. Republication or reproduction for sale of pages in this book is permitted only under license from ACS. Direct these and other permission requests to ACS Copyright Office, Publications Division, 1155 16th St., N.W., Washington, DC 20036.

The citation of trade names and/or names of manufacturers in this publication is not to be construed as an endorsement or as approval by ACS of the commercial products or services referenced herein; nor should the mere reference herein to any drawing, specification, chemical process, or other data be regarded as a license or as a conveyance of any right or permission to the holder, reader, or any other person or corporation, to manufacture, reproduce, use, or sell any patented invention or copyrighted work that may in any way be related thereto. Registered names, trademarks, etc., used in this publication, even without specific indication thereof, are not to be considered unprotected by law.

PRINTED IN THE UNITED STATES OF AMERICA

Foreword

The ACS Symposium Series was first published in 1974 to provide a mechanism for publishing symposia quickly in book form. The purpose of the series is to publish timely, comprehensive books developed from ACS sponsored symposia based on current scientific research. Occasionally, books are developed from symposia sponsored by other organizations when the topic is of keen interest to the chemistry audience.

Before agreeing to publish a book, the proposed table of contents is reviewed for appropriate and comprehensive coverage and for interest to the audience. Some papers may be excluded to better focus the book; others may be added to provide comprehensiveness. When appropriate, overview or introductory chapters are added. Drafts of chapters are peer-reviewed prior to final acceptance or rejection, and manuscripts are prepared in camera-ready format.

As a rule, only original research papers and original review papers are included in the volumes. Verbatim reproductions of previously published papers are not accepted.

ACS Books Department

Contents

Preface..xi

Background

1. Polymer Processing in Microgravity: An Overview........................2
   James Patton Downey and John A. Pojman

2. Research Platforms That Reduce the Acceleration
   Experienced in the Experimental Frame of Reference:
   Achieving "Microgravity"..16
   James Patton Downey

Sounding Rocket and Orbital Investigations

3. The Role of Gravity in Sol–Gel Systems......................................32
   Laurent Sibille and David D. Smith

4. Growth Kinetics of Polydiacetylene Films Prepared
   in Microgravity..51
   William E. Carswell, Mark S. Paley, Donald O. Frazier,
   and Robert J. Naumann

5. Polymerization in Microgravity On-Board STS–57,
   STS–63, and STS–77...64
   Kenneth G. Brown, Karen S. Burns, Jo Ingram,
   George M. Wood, and Billy T. Upchurch

6. Production of Polyurethane Foams in Space:
   Gravitational and Vacuum Effects on Foam Formation...............78
   Samuel P. McManus, Francis C. Wessling, John T. Matthews,
   Darayas N. Patek, Kazunori Emoto, Douglas T. Howard,
   and T. Scott Price

7. Gel Formation under Microgravity Conditions..........................97
   V. A. Briskman and K. G. Kostarev

Parabolic Flight Investigations

8. Bubble Behavior in Frontal Polymerization:
   Results from KC–135 Parabolic Flights................................112
   William J. Ainsworth, John A. Pojman, Yuri A. Chekanov,
   and Jonathan Masere

9. Instrumentation for Studying Polymer Film
   Formation in Low Gravity...126
   Paul Todd, Matthew R. Pekny, Jeremiah Zartman,
   William B. Krantz, and Alan R. Greenberg

10. Liquid Crystal and Polymer Dispersions
    in a Microgravity Environment..138
    Joe B. Whitehead, Jr., and Gregory P. Crawford

11. Processing of Nonlinear Optical Polymer Systems
    under Microgravity Conditions..153
    L. H. Dao, K. Parbhakar, H. M. Nguyen, J. M. Yin, Y. Sun,
    and Y. Beaudoin

Ground-Based Research

12. Foaming of Polyethylenes in a Dynamic Decompression
    and Cooling Process..168
    Kwangjin Song and Robert E. Apfel

13. Prospects for the Study of Gas-Phase Polymerization
    and the Synthesis of Polymers Containing Nanoparticles
    in Microgravity..185
    Yezdi B. Pithawalla, Junling Gao, and M. Samy El Shall

14. Production of Bacterial Polyesters
    in Simulated Microgravity...203
    Radhika Thiruvenkatam and Carmen Scholz

15. Development of Impedance Spectroscopy to Study
    Polymerization Processes in Microgravity.................................217
    A. P. Kennedy and J. McLendon

16. Gravity Effects during Mold Filling: Fluid Front Dynamics..........227
    M. C. Altan and K. A. Olivero

Author Index..245

Subject Index...246

Preface

Research is frequently hampered by effects associated with the force of gravity. Examples include the sedimentation of colloids, the non-uniformity of pressure in fluids due to the pressure head, the difficulties of studying surface-tension driven convection, or diffusive phenomena due to the presence of buoyancy-driven convection. All are relevant to polymeric research.

Because gravity arises from a basic property of matter, it is not an annoyance that an investigator can simply eliminate. However, research facilities exist in which conditions close to "free fall" can be achieved. "Free fall" is a condition such that all objects in the laboratory reference frame are allowed to accelerate at a rate consistent with the gravitational attraction. If all objects and fluid bodies in the experimental reference frame accelerate in a manner consistent with gravity, there is no sedimentation, buoyancy-driven convection, or pressure head observed within the experimental reference frame. This is referred to as "weigthlessness". The term microgravity has been coined to describe these conditions since at best the average deviation from uniform acceleration consistent with gravity is on the order of one part in one million.

Drop towers, parabolic airplane trajectories, and spacecraft have all been utilized to achieve microgravity conditions. Within the United States, the National Aeronautics and Space Administration (NASA) is responsible for maintaining, utilizing, and financing research in these facilities. At the time of publication, the construction of the International Space Station was well underway. This facility will provide greater access to microgravity laboratory conditions than any previous space facility. As a result the magnitude of microgravity research is expected to increase significantly in the near future.

Microgravity studies have been conducted since the days of the Apollo and Skylab programs, approximately 30 years ago. However, due to the expensive and infrequent nature of space flights the number of experiments has been relatively small considering the number of years that have passed. Relatively few microgravity experiments have been performed involving polymeric materials. Even considering the difficulties associated with such experiments this seems odd. One partial explanation for this is that the Space Studies Board of the National Research Council advised NASA, "...the viscous character of most high polymer melts greatly desensitizes their response to gravitational acceleration..." (1). This statement has perhaps been interpreted to mean that relatively few microgravity experiments involving polymers are of interest.

However, most polymer processes and polymerizations are initiated in a relatively low-viscosity state. Much polymer work is performed in dilute solutions with viscosities similar to that of pure water. In fact, polymeric research spans a spectrum of viscosity levels—often within a single process.

Few review articles on microgravity research exist and none to the authors' knowledge are specific to macromolecular research. Considering the imminent start of Space Station research and the importance of polymers in modern life, now is the appropriate time to review the history of polymeric microgravity research, to discuss polymeric systems most affected by gravity, and to speculate on the future of microgravity polymeric research.

The primary purpose of this American Chemical Society (ACS) Symposium Series publication is to review previous macromolecular research that has been performed using a microgravity environment, to provide a forum for current work, and in a broader sense to review topics that are appropriate for study in this environment. This volume is based on presentations made in the symposium "Polymer Processing in Microgravity" held at the 219th National Meeting of the ACS held March 26–31, 2000 in San Francisco. This was not the first symposium related to the topic. In 1987, Sam McManus, an author in this volume, organized a symposium entitled "Microgravity Research & Processing," which included among many topics in materials science, polymers. It was held at the National Meeting of the ACS in New Orleans (2). Nonetheless, to our knowledge, this is the first book dedicated to polymerization and polymer processing in microgravity.

Previous research involving the effect of gravity on macromolecules has lead to a number of interesting results most of which are covered in the introductory chapter of this volume. The second chapter provides an overview of the types of microgravity facilities available. Later chapters provide detailed accounts of individual research efforts associated with avoiding and/or understanding gravitational effects in the field of macromolecular science. Polymerization reactions and the processing of polymers in microgravity are included. Research from both ground-based and flight-based investigations is presented. The entire range of polymers is spanned from biological production of polymers to inorganic sol–gel systems.

One topic not covered here is the extensive effort in crystallizing biological macromolecules and complexes in microgravity. An interested reader may wish to read the review by McPherson on the subject (3).

References

1. Microgravity Resource Opportunities for the 1990's; Library of Congress Catalog Number: 94–74623; National Academy oc Sciences: Washington, DC, 1995.

2. *Polymer Preprints* **1987**, *26,* 451–464
3. McPherson, A. *Crystallogr. Rev.* **1996,** *6,* 157–308.

James Patton Downey
Microgravity Department
George C. Marshall Space Flight Center
SD48
National Aeronautics and Space Administration
Huntsville, AL 35812

John A. Pojman
Department of Chemistry and Biochemistry
University of Southern Mississippi
Hattiesburg, MS 39406

Background

Chapter 1

Polymer Processing in Microgravity: An Overview

James Patton Downey[1] and John A. Pojman[2]

[1]Microgravity Department, George C. Marshall Space Flight Center, SD48, National Aeronautics and Space Administration, Huntsville, AL 35812
[2]Department of Chemistry and Biochemistry, University of Southern Mississippi, Hattiesburg, MS 39406

Understanding how polymer processing can be different in microgravity requires an understanding of how gravity can affect polymer reactions and processes. We review here the fundamentals of buoyancy-driven (Rayleigh-Bénard) and surface-tension driven (Marangoni) convection. We consider the polymer processes that are affected by convection and review polymer experiments that demonstrate convective effects in 1 g and microgravity.

Why should gravity be an issue for polymer processing, or for that matter, any chemical process? A simple calculation seems to suggest that gravity should have negligible influence on chemical reactions. The mass of a molecule is on the order of 10^{-26} kg, which translates into a gravitational force of about 10^{-25} N. We can compare this to the force of attraction between the electron and proton in a hydrogen atom, which is of the order 10^{-8} N. Even allowing for shielding effects, the electrostatic forces that cause chemical bonds to be made and broken will always be many orders of magnitude stronger than gravitational forces. So gravity does not affect the fundamental atomic and molecular interactions, but it can drastically alter the macroscopic transport of heat and matter through *convection*, or macroscopic fluid motion. Gravity is also very

© 2001 American Chemical Society

important for systems with different phases that can sediment, as in colloids, emulsions and blends.

Natural convection is the movement of fluid as the result of differences in density, so that denser fluid sinks and less dense fluid rises. This motion is resisted by the viscosity of the medium, which acts like friction in slowing the motion of solids. The study of convection is an entire area of physics, and we will touch only on a few aspects. The reader is referred to some excellent texts (*1-3*).

Convection is a much more efficient process than diffusion for transporting heat and matter. Consider what would happen if smoke in a fireplace were removed solely by diffusion. In a short time, the room would fill as the smoke particles dispersed randomly. Instead, if things work properly, the smoke goes up the chimney, as the exothermic combustion reactions in the fire produce heat, which decreases the density of the gases and allows them to "float" up the flue because of buoyancy. We understand this idea of buoyancy intuitively when we say that "heat rises".

Gravity will not always cause convection - it will depend on several factors including the viscosity, temperature and concentration gradients, thermal and mass diffusivities and the orientation of gradients with respect to the gravity vector. We understand this intuitively when thinking of hot water over cold water being stable. (We'll see that hot salty water over cold water can still exhibit convection through a more complicated mechanism.)

If a system lacks an interface between different fluids, such a monomer/air interface, then only buoyancy-driven convection will occur. If a free interface exists, then we will see that gradients in the interfacial tension can cause fluid motion -- a process called Surface-Tension Induced Convection or Marangoni convection. This will be especially important in "microgravity". (How the condition of apparently zero gravity is achieved is discussed in chapter 2.)

Buoyancy-Driven Convection

Consider a single-component fluid in a rectangular container with no free interface, as shown in Figure 1. We can heat it from the top, bottom or side. Will all three configurations cause fluid motion? No. If we heat from above, the fluid is stable, meaning no fluid motion will occur. If we heat from the side, there will be convection, whose magnitude will be determined by several factors. We combine these factors into a dimensionless quantity called the Rayleigh number

$$Ra = \frac{g\alpha\Delta T d^3}{\kappa \nu}$$

where α is the thermal expansion coefficient, $(\partial\rho/\partial T)/\rho_0$, d is the height of the container, κ the thermal diffusivity (cm^2/s), and ν the kinematic viscosity (cm^2/s) and g the gravitational acceleration. Convection is inevitable in any fluid system where there are temperature gradients perpendicular to the gravity vector and the Rayleigh number is non-zero.

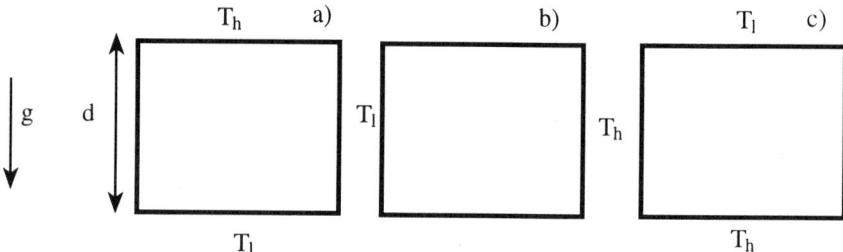

Figure 1. Schematic of heating a single-component fluid. a) heaingt from above is stable. b) heating from the side always causes convection. c) heating from below may cause convection

What about heating from below? The answer is that it depends. If we have an infinite layer with a fixed boundaries above and below, then the critical Rayleigh number is 1707 (3). Such convection is often called Bénard convection or Rayleigh-Bénard convection, in tribute to Henri Bénard who first investigated the problem (4). If heating from below, the Rayleigh number must be below a critical value specific to the geometry of the system or convection will occur. Another important observation is that to reduce the Rayleigh number to near zero without altering the gradients, we can reduce "g" OR increase the viscosity. Adding something like silica gel (5) or turning the reactor over (6) can answer important information without resorting to microgravity.

We note that convection can also be caused by concentration gradient. In that case, the αΔT term in the Rayleigh number is replaced by the βΔC, where β = $(\partial\rho/\partial C)/\rho_0$, and κ is replaced by the diffusion coefficient.

$$Ra = \frac{g\beta\Delta C d^3}{D\nu}$$

This has an important implication, especially for polymer systems. The diffusion coefficient of typical solute solvent systems is on the order of 10^{-5} cm^2

s^{-1} and may be orders of magnitude smaller for macromolecules. Thus, a density gradient several orders smaller than a comparable thermally-induced gradient will still cause convection.

The situation becomes more complicated if concentration and temperature gradients occur simultaneously. For example, if hot, salty water overlies cold, fresh water, vigorous convection will occur even if the net density gradient is zero. This phenomenon is called "double-diffusive convection" (*2,7,8*). The appearance of "salt fingers" is quite obvious is such cases (*9*). If the hot, salty water underlies cold fresh water, then the "diffusive regime" occurs, with less vigorous convection for the same magnitude of gradients as for the "salt finger" regime.

For the double-diffusive instability to occur two components with differing effects on the density must be present that have transport coefficients that differ by at least a factor of three (*2*). If temperature and concentration gradients occur, this condition is readily fulfilled because thermal diffusivities of liquids like water and organic fluids are on the order of 10^{-3} cm^2 s^{-1}. It also can be fulfilled by a polymer solution over a salt solution even though the polymer solution has a lower density than the salt solution due to the large differences in the diffusivities of small and large molecules (*10,11*). In a reactive polymeric system the system is self-heated with a simultaneous chemical change making the stability more complicated than for the heating of a single-component fluid; double- diffusive convection may result.

Another instability can occur when a more dense fluid is placed on top of a less dense fluid, called the Rayleigh-Taylor instability (*12,13*). This can be observed with a hot thermoplastic over its monomer (*5*).

We should note that if an initially homogeneous solution is heated, concentration gradients will arise from the Soret effect (*14*). This is due to the temperature dependence of the solute's chemical potential. For a polymer solution in a thermal gradient, the polymer molecules may migrate toward the lower temperature region. The Soret effect is usually small. However, the Soret coefficient is typically 100 times larger for macromolecules than for small molecules (*15,16*).

Surface-Tension Driven Convection

Even in the absence of buoyant forces, convection can occur, driven by gradients in the interfacial (surface) tension at the interface of two fluids. Surface tension is affected both by chemical concentration and by temperature. Figure 2 shows how a hot spot can cause convection by locally lowering the surface tension. The cooler fluid has a higher surface tension and draws the warm fluid towards itself. If the temperature gradient is perpendicular to the interface, both buoyancy-driven convection and Marangoni (surface-tension driven) convection are possible.

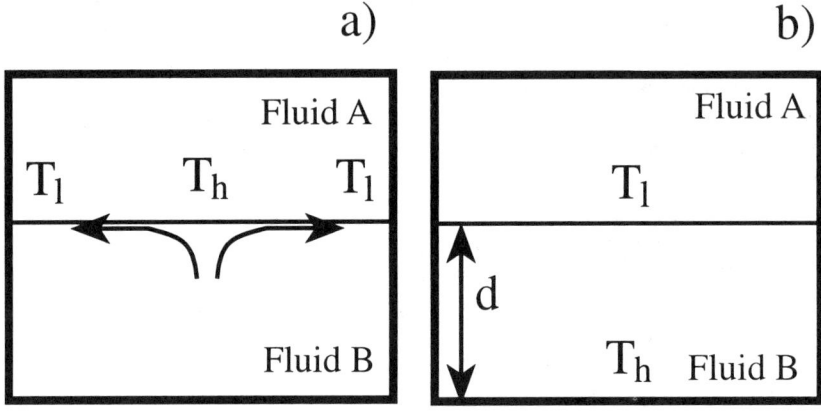

Figure 2. a) A temperature gradient along an interface between two fluids causes a gradient in surface tension, which always results in fluid motion. b) A difference in temperature between the interface and the bottom of the container may cause convection, depending on the temperature difference, the transport coefficients and the depth of the container.

The Marangoni number is the analog to the Rayleigh number but in which the driving force is the variation in the interfacial tension,

$$Ma = -\frac{\partial \sigma}{\partial T}\frac{\Delta T d}{\mu \kappa}$$

where μ is the dynamic viscosity, ν ρ, and $\partial\sigma/\partial T$ has units of N m^{-1} K^{-1}. As in the buoyancy-driven case, any lateral temperature (or concentration gradient) will cause convection whose magnitude will be determined by the value of the Marangoni number. If the gradient is vertical, then a stability condition exists, i. e., below a critical value of the Marangoni number no convection occurs ([17]). We hasten to add that the interfacial tension can be a function of concentration so that a solutal Marangoni number is defined analogously to the solutal Rayleigh number.

In a one g environment, both buoyancy-driven convection and Marangoni convection can occur in a system with a free interface, depending on the orientation of the gradient and the values of Ma and Ra. With thermal or concentration gradients parallel to the interface, both effects occur. Separating them can be accomplished by reducing g to 0 because the Ma number does not depend on g. Increasing the viscosity will not work because both G and M depend inversely on the viscosity. Another way to separate the influence of the two types of convection is to vary d. The Marangoni number depends on the first power of the layer depth while the Rayleigh number, which describes bulk convection, depends on d^3. Hence, surface tension-driven convection dominates in thin layers.

Polymer Systems Affected by Gravity

There is a misconception that polymers are not affected by convective instabilities. Consider a common material like poly(methyl methacrylate), aka Plexiglass. At room temperature it is a rigid solid. At about 100 °C it will soften. At higher temperatures it will flow. Depending on its molecular weight, at high temperature it can have a viscosity similar to water. During the processing of thermoplastics, convection can be important. More importantly, the viscosity of the monomer is low, and it is only after high conversion is achieved that the viscosity becomes high.

There are several reasons to study polymer processing in microgravity, which can be grouped into three categories. The first of which is the principal concern of this review and the majority of microgravity experiments. Namely, microgravity greatly reduces buoyancy-driven fluid flow. This allows the study of Marangoni effects on processes without most of the convoluting influence of buoyancy-driven convection. It can also allow researchers to measure diffusion coefficients in experiments where heat and/or mass transfer are normally dominated by buoyant convection. Finally, the near absence of buoyant-convection allows the formation of benchmark materials. These benchmarks allow the characterization of materials formed under conditions where buoyancy-driven convective influences are insignificant. Such benchmarks are of interest for determining the influence of buoyancy-driven convection on materials properties.

A second useful feature of microgravity is that pressure head effects are virtually eliminated. This is important in phase change studies where small gradients in pressure cause significant changes in the behavior of the phases.

Third, microgravity virtually eliminates sedimentation. In emulsion polymerization and dispersion polymerization polymer particles are formed with a higher density than the surrounding medium and tend to settle. The same sedimentation issues arise with polymer production in living cells. Microgravity allows a researcher to create colloids and dispersions with stabilities not normally achievable on the ground.

The following pages are a brief review of experiments indicative of gravitational effects on polymerizations and polymer processes. These experiments provide an insight into the types of processes that may be investigated using microgravity experiments.

Bulk Polymerizations

Even crosslinking systems can be affected by buoyancy-driven convection because the viscosity of the reactants are often quite low. This is certainly true for acrylates and often true for epoxy resins. Li et al. demonstrated that the Rayleigh-Bénard patterns that occur in a monomer system heated from below can be fixed by the polymerization (18). Orbán et al. studied pattern formation during the polymerization of acrylamide in water in the presence of sulfide ions and found that that convection caused the pattern and not the supposed Turing mechanism (19,20). Schaarschmidt and Lamprecht investigated two systems in thin layers (1 - 5 mm): the polymerization of acrylamide and the gelatinization of gelatin (21). From their experiments and analysis they concluded it was impossible to determine whether buoyancy-driven or surface-tension driven convection was more important. Briskman et al. studied the photopolymerization of acrylamide in water on the *Mir* (22,23). Briskman et al. also studied the same system under high g (24) and one g (25). They found that the gel matrix properties were graded along the direction of the force.

The effects of convection during the photodeposition of poly(diacetylene) thin films has been studied at Marshall Space Flight Center (26,27). The investigators found that defect formation was related to the intensity of convection and that double-diffusive convection and defects were found in all orientations of the experiment including illumination from above.

Buoyancy-driven convection can be important in the dissolution of polymers (28,29). This process is the reverse of solidification but shares the trait of concentration gradients created near a solid surface. A polymer process closely related to solidification is isothermal frontal polymerization in which an acrylate polymerization occurs in the mushy layer of a polymer "seed" (30). This process can be used to create Gradient Refractive Index (GRIN) fibers (31). Koike et al. also developed a photopolymerization method (32). Liu et al. used a centrifugal method to prepare the gradients (33). Because of the large conversion gradients created, the orientation of the reactor is an important processing variable.

Mathias et al. studied the photopolymerization of methyl methacrylate on parabolic airplane flights and reported a variation in the molecular weight with g level (34) consistent with a report from a drop tower experiment by Sturm et al. (35). This phenomenon is most likely an effect of mixing caused by composition gradients created during photopolymerization that affects the average rate of initiation throughout the sample.

On the space shuttle flights STS-57 and STS-63 Brown et al. studied polymer gel formation and produced superior gels for gas separation and for proprietary applications (*36,37*).

Coatings, Films and Membranes

Coatings and films involve the creating of density and surface-tension gradients. Hansen and Pierce reviewed the appearance of cellular patterns in coatings and its deleterious effects on coating appearance (*38*). They note that this convection is caused primarily by gradients of surface tension due to the small scale of the layer. Kitano and Shiojiri used this convection to create a superior electrostatic dissipative coating with ZnO particles formed into a network by cellular convection (*39*). Anand and Karam studied the effects of surface-tension induced convection on surface deformation in Saran coating during the evaporation of the solvent (*40*). Sakurai et al. studied pattern formation in polymer films formed by evaporating toluene from a solution containing poly(styrene-*ran*-butadiene) random copolymer and from toluene solution containing polystyrene and dioctylphthalate) (*41*). They attributed the hexagonal cellular patterns to buoyancy-driven convection but neglected any consideration of the effect of surface-tension driven flows.

McGinniss and Whitmore studied thin polysulfone membranes formed under high and low g environments on parabolic airplane flights, sounding rockets and on the Space Shuttle (*42*). The most important effect of acceleration level was on the solvent gas transfer coefficient and not on the components in the membrane. Tan et al. considered the same issue theoretically (*43*).

Kawaguchi studied the problem of viscous fingering in silica suspensions in polymer solutions (*44*). Experiments in drop towers revealed that sedimentation of the silica particles affected slightly the tip-splitting instability.

Phase Separation

Phase separating polymer systems can exhibit both buoyancy-driven and Marangoni convection. Mitov and Kumacheva observed both in the polystyrene/poly(methyl methacrylate) system in which both were dissolved in toluene. The toluene was evaporated from layers of initial thickness of 0.12 - 3 mm (*45*). They were able to obtain ordered hexagonal patterns on the scale of 10 microns, which they ascribed to the ordering effects of buoyancy-driven and Marangoni convection.

Kumaki et al. studied the polystyrene-polybutadiene-dioctylphthalate system 20 °C above the coexistence temperature but in the presence of a temperature gradient. Spinodal-like patterns were observed, which were caused by surface-tension-induced convection (46).

Polymer-Dispersed Liquid Crystals (PDLCs) can be affected by buoyancy. Fox et al. studied PDLCs produced via photopolymerization on parabolic airplane flights (47). The authors reported improved response times in the flight samples.

Li et al. (48) studied how 1.0 µm particles of polystyrene and poly(methyl methacrylate) interacted when they were melt processed at 180 °C. They observed by confocal microscopy on a hot stage that there was a preferential motion for particles, which they attributed to a buoyancy-driven flow because of the 10% density difference between the polymers. Jang et al. had made a similar observation (49). Li et al. did not consider possible surface-tension induced convection or that droplets could migrate in a temperature gradient, as has been observed by Balasubramaniam et al. for the thermocapillary migration of bubbles (50).

Naumann reviewed the results from the *Consort* sounding rocket program that included several polymer experiments but to the best of the authors' knowledge none of experiments have been published in the peer-reviewed literature (51).

Sunkara et el. studied the effects of buoyancy-driven convection on colloidal crystals consisting of polystyrene beads in a photopolymerizing methyl methacrylate matrix (52). Buoyancy-driven convection occurred no matter in which orientation the UV light source was placed. This convection prevented fixing a uniform colloid matrix by polymerization.

Vanderhoff et al. prepared large-particle size monodisperse latexes on the Space Shuttle (53). These latexes were made from large colloidal droplets consisting primarily of monomers that were polymerized. Droplets of similar size could not be maintained on the ground for sufficient duration due to sedimentation.

Rabeony and Weiss investigated theoretically nonterminated polymerization in the gas phase without convection (54).

Tubular Reactors

Tubular reactors have been shown to be affected by buoyancy-driven convection. Cunningham et al. demonstrated the fouling of a reactor (when polymer accumulated on the reactor walls and stopped flow) depended on the reactor orientation (55-57).

Foams

Foams are quite susceptible to influences from the buoyancy. Polymeric foam formation has been studied on sounding rockets (58,59) and parabolic airplane flights (60,61). Volatilization of thermoplastics is also affected, which affects how thermoplastics can burn (62).

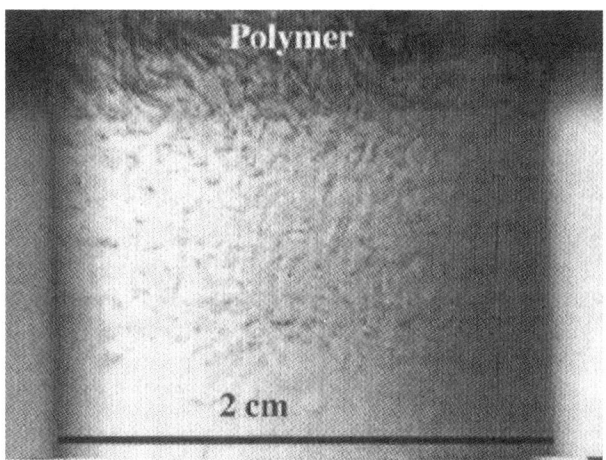

Figure 3. Rayleigh-Taylor instability with descending front of butyl acrylate polymerization. Although the polymer product is hot (> 200 °C) it still is about 20% more dense than the monomer below it.

Frontal polymerization

Frontal polymerization is a mode of converting monomer into polymer via a localized exothermic reaction zone that propagates through the coupling of thermal diffusion and Arrhenius reaction kinetics. Frontal polymerization was discovered in Russia by Chechilo and Enikolopyan in 1972 (63). The macrokinetics and dynamics of frontal polymerization have been examined in detail (5) and applications for materials synthesis considered (64,65).

Buoyancy-driven convection has significant effects on ascending fronts, with thermosets (6,66) and thermoplastics (67). It has been used to advantage for creating functionally-gradient materials (68,69).

Descending fronts with thermosets are generally immune to convective instabilities but thermoplastics exhibit the Rayleigh-Taylor instability (Figure 3)

(5). However, if the reactor is not vertical, i.e., the cylindrical tube is not parallel to the gravitational vector, then convection will occur with a thermoset (70). Descending fronts are affected by gravity even without convective instabilities. Under some conditions and/or composition, the fronts propagate in a spin mode forming a spiral pattern (5,71,72). Garbey et al. had predicted (73) and Masere et al. confirmed experimentally (72) that for descending fronts spin modes appeared more readily for high Rayleigh numbers than for low values.

In order to study poly(n-butyl acrylate) fronts, Pojman et al. added ultrafine silica gel to increase the monomer viscosity (74). To determine the inherent molecular weight distribution (MWD) of poly(n-butyl acrylate) produced frontally, it was necessary to perform a front without silica gel. Because of the Rayleigh-Taylor instability, the only way to accomplish this with a thermoplastic-forming monomer such as n-butyl acrylate is to eliminate the driving force for the collapse of the more dense polymer layer, i.e., eliminate the force of gravity. An experiment was launched on the 1996 *Conquest I* rocket flight. The flight provided at least 5 minutes of 10^{-4} g conditions. Pojman et al. found that the molecular weight distribution of the sample polymerized in microgravity was very similar to the ground based-control experiment (75). Thus, the addition of a viscosity-enhancing agent did not significantly affect the molecular properties of the sample produced frontally.

Conclusions

Buoyancy-driven and surface-tension driven convection can affect a wide variety of polymer process. Often the role can be studied on earth by varying the viscosity or orientation or the system. Yet, performing experiments in weightlessness can be the only way to determine the relative effects of the two processes or if the viscosity can not be independently varied.

References

1. Tritton, D. J. *Physical Fluid Dynamics*; Oxford University Press: Oxford, 1988.
2. Turner, J. S. *Buoyancy Effects in Fluids*; Cambridge University Press: Cambridge, 1979.
3. Antar, B. N.; Nuotio-Antar, V. S. *Fundamentals of Low Gravity Fluid Dynamics and Heat Transfer*; CRC Press: Boca Raton, 1993.
4. Bénard, H. *Ann. Chim. Phys.* **1901**, *23*, 62-?

5. Pojman, J. A.; Ilyashenko, V. M.; Khan, A. M. *J. Chem. Soc. Faraday Trans.* **1996**, *92*, 2825-2837.
6. Bowden, G.; Garbey, M.; Ilyashenko, V. M.; Pojman, J. A.; Solovyov, S.; Taik, A.; Volpert, V. *J. Phys. Chem. B* **1997**, *101*, 678-686.
7. Turner, J. S. *Annu. Rev. Fluid Mech.* **1985**, *7*, 11-44.
8. Pojman, J. A.; Komlósi, A.; Nagy, I. P. *J. Phys. Chem.* **1996**, *100*, 16209-16212.
9. Nagy, I. P.; Pojman, J. A. *J. Phys. Chem.* **1993**, *97*, 3443-3449.
10. Nason, P.; Schumaker, V.; Halsall, B.; Schwedes, J. *Biopolymers* **1969**, *7*, 241-249.
11. Comper, W. D.; Preston, B. N. *J. Coll. Interface Sci.* **1984**, *99*, 305-314.
12. Rayleigh, J. W. *Scientific Papers, ii*; Cambridge University Press: Cambridge, 1899.
13. Taylor, G. *Proc. Roy. Soc. (London)* **1950**, *Ser. A. 202*, 192-196.
14. de Groot, S. R.; Mazur, P. *Nonequilibrium Thermodynamics*; Dover: New York, 1984.
15. Schimpf, M. E.; Giddings, J. C. *J. Polym. Sci. Polym. Phys. Ed.* **1989**, *27*, 1317-?
16. Giddings, J. C. *Science* **1993**, *260*, 1456-?
17. Sternling, C. V.; Scriven, L. E. *A.I.Ch.E. J.* **1959**, *5*, 514-523.
18. Li, M.; Xu, S.; Kumacheva, E. *Macromolecules* **2000**, *33*, 4972-4978.
19. Kurin-Csörgei, K.; Orbán, M.; Zhabotinsky, A. M.; Epstein, I. R. *Chem. Phys. Ltts.* **1998**, *295*, 70-74.
20. Orbán, M.; Kurin-Csörgei, K.; Zhabotinsky, A. M.; Epstein, I. R. *J. Phys. Chem. B.* **1999**, *103*, 36-40.
21. Schaarschmidt, B.; Lamprecht, I. *J. Therm. Anal.* **1991**, *37*, 1833-1839.
22. Briskman, V. A.; Kostarev, K. G. *Polym. Prepr. (Am Chem. Soc. Div. Polym. Chem.)* **2000**, *41*, 1054-1055.
23. Righetti, P. G.; Bossi, A.; Giglio, M.; Vailati, A.; Lyubimova, T.; Briskman, V. A. *Electrophoresis* **1994**, *15*, 1005-1013.
24. Briskman, V. A.; Kostarev, K. G.; Levtov, V.; Lyubimova, T.; Mashinsky, A.; Nechitailo, G.; Romanov, V. *Acta Astronautica* **1996**, *39*, 395-402.
25. Briskman, V. A. *Adv. Space Res.* **1999**, *24*, 1199-1210.
26. Paley, M. S.; Frazier, D. O.; Abdeldeyem, H.; Armstrong, S.; McManus, S. P. *J. Am. Chem. Soc.* **1995**, *117*, 4775-4780.
27. Frazier, D. O.; Hung, R. J.; Paley, M. S.; Long, Y. T. *J. Cryst. Growth* **1997**, *173*, 172-181.
28. Sarti, G. C.; Gosstoli, C.; Riccioli, G.; Carbonell, R. G. *J. Appl. Polym. Sci.* **1986**, *32*, 3627-2647.
29. Ranade, V. V.; Mashelkar, R. A. *AIChE J.* **1995**, *41*, 666-675.
30. Smirnov, B. R. M. k., S. S.; Lusinov, I. A.; Sidorenko, A. A.;Stegno, E. V.; Ivanov, V. V. *Vysokomol. Soedin., Ser. B* **1993**, *35*, 161-162.

31. Koike, Y. Gradient Index Materials and Components: in *Polymers for Lightwave and Integrated Optics*; Hornak, L. A., Ed.; Marcel Dekker: New York, 1992; pp 71-104.
32. Koike, Y.; Hatanaka, H.; Otsuka, Y. *Appl. Opt.* **1984**, *23*, 1779-1783.
33. Liu, J.-H.; Chen, J.-L.; Wang, H.-Y.; Tsai, F.-R. *Macromolecular Chemistry and Physics* **2000**, *201*, 126-131.
34. Avci, D.; Thigpen, K.; Mathais, L. J. *Polym. Prepr. (Am Chem. Soc. Div. Polym. Chem.)* **1996**, *37(2)*, 317.
35. Sturm, D.; Müller, R.; Rath, M.-J. Photoinitiated Radical Polymerization of Liquid Monomers in Microgravity, *VIIIth European Symposium on Materials and Fluid Sciences in Microgravity,* **1992**, 895-899.
36. Brown, K. G.; Burns, K. S.; Upchurch, B. T.; Wood, G. *Proc. ACS; Div. Poly. Mater. Sci. & Eng.* **1996**, *74*, 296-297.
37. Brown, K. G.; Burns, K. S.; Upchurch, B. T.; Wood, G. M. *Polym. Prepr. (Am Chem. Soc. Div. Polym. Chem.)* **2000**, *41*, 1072-1073.
38. Hansen, C. M.; Pierce, P. E. *Ind. Eng. Chem. Prod. Res. Develop.* **1973**, *12*, 67-70.
39. Kitano, M.; Shiojiri, M. *Powder Technology* **1997**, *93*, 267-273.
40. Anand, J. N.; Karam, H. J. *J. Coll. Interface Sci.* **1969**, *31*, 208-215.
41. Sakurai, S.; Tanaka, K.; Nomura, S. *Polymer* **1993**, *34*, 1089-1092.
42. McGinniss, V. D.; Whitmore, R. S. *J. Mem. Sci.* **1995**, *98*, 27-47.
43. Tan, L.; Krantz, W. B.; Greenberg, A. R.; Sani, R. L. *J. Membrane Sci.* **1995**, *108*, 245-255.
44. Kawaguchi, M.; Makino, K.; Kato, T. *Physica A* **1997**, *246*, 385-398.
45. Mitov, Z.; Kumacheva, E. *Phys. Rev. Ltts.* **1998**, *81*, 3427-3430.
46. Kumaki, J.; Hashimoto, T.; Grannick, S. *Phys. Rev. Ltts.* **1996**, *77*, 1990-1993.
47. Fox, B. H.; Schuster, P. A.; Kilp, T.; Fuh, A. Y. G. A Study of Polymer Dispersed Liquid Crystals (PDLC) Fabricatd in Both Microgravity and Terrestrial Conditions, *AIAA 93-0578,* **1993**,
48. Li, L.; Sosnowski, S.; Kumacheva, E.; Winnik, M. W. *Langmuir* **1996**, *12*, 2141-2144.
49. Jang, B. Z.; Uhlmann, D. R.; Vander Sande, J. B. *Rubber Chem. Technol.* **1984**, *57*, 291-?.
50. Balasubramaniam, R.; Lacy, C. E.; Woniak, G.; Subramanian, R. S. *Phys. Fluids* **1996**, *8*, 872-880.
51. Naumann, R. J. *Microgravity Sci. Technol.* **1995**, *VIII*, 204-213.
52. Sunkara, H. B.; Penn, B. G.; Frazier, D. O.; Ramachandran, N. *J. Mater. Sci.* **1998**, *33*, 887-894.
53. Vanderhoff, J. W.; El-Aasser, M. S.; Micale, F. J.; Sudol, E. D.; Tseng, C.-M.; Silwanowicz, A.; Kornfeld, D. M. *J. Dispersion Sci. Technology* **1984**, *5*, 231-246.
54. Rabeony, H.; Reiss, H. *Macromolecules* **1988**, *21*, 912-918.

55. Cunningham, M. F.; O'Driscoll, K. F.; Mahabadi, H. K. *Can. J. Chem. Eng.* **1991**, *69*, 630-638.
56. Cunningham, M. F.; O'Driscoll, K. F.; Mahabadi, H. K. *Polym. Reaction Eng.* **1992**, *1*, 229-244.
57. Cunningham, M. F.; O'Driscoll, K. F.; Mahabadi, H. K. *Polym. Reaction Eng.* **1992**, *1*, 245-288.
58. Wessling, F. C.; McManus, S. P.; Matthews, J.; Patel, D. *J. Spacecraft* **1990**, *27*, 324-329.
59. McManus, S. P. *Polym. Prepr. (Am Chem. Soc. Div. Polym. Chem.)* **2000**, *41*, 1056-1057.
60. Curtin, M. M.; Tyler, F. S.; Wilkinson, D. *J. Cell. Plastics* **1992**, *28*, 536-556.
61. Noever, D. A. *J. Spacecraft and Rockets* **1994**, *31 No. 2*, 319-322.
62. Wichman, I. S. *Combust. Flame* **1986**, *63*, 217-229.
63. Chechilo, N. M.; Khvilivitskii, R. J.; Enikolopyan, N. S. *Dokl. Akad. Nauk SSSR* **1972**, *204*, 1180-1181.
64. Khan, A. M.; Pojman, J. A. *Trends Polym. Sci. (Cambridge, U.K.)* **1996**, *4*, 253-257.
65. Pojman, J. A.; Fortenberry, D.; Lewis, L. L.; Simmons, C.; Ilyashenko, V. *ACS Symp. Ser. (Solvent-Free Polymerization and Processes)* **1998**, *713*, 140-153.
66. Chekanov, Y.; Arrington, D.; Brust, G.; Pojman, J. A. *J. Appl. Polym. Sci.* **1997**, *66*, 1209-1216.
67. McCaughey, B.; Pojman, J. A.; Simmons, C.; Volpert, V. A. *Chaos* **1998**, *8*, 520-529.
68. Pojman, J. A.; Chekanov, Y. A.; Case, C.; McCardle, T. *Polym. Mater. Sci. Eng. Prep. Div. Polym. Mater. Sci. Eng.* **1998**, *79*, 82-83.
69. Pojman, J. A.; McCardle, T. W. U.S. Patent 6,067,406,2000
70. Pojman, J. A.; Nichols, H. A.; Volpert, V. *Polym. Prepr. (Am Chem. Soc. Div. Polym. Chem.)* **1998**, *39*, 463-464.
71. Pojman, J. A.; Ilyashenko, V. M.; Khan, A. M. *Physica D* **1995**, *84*, 260-268.
72. Masere, J.; Stewart, F.; Meehan, T.; Pojman, J. A. *Chaos* **1999**, *9*, 315-322.
73. Garbey, M.; Taik, A.; Volpert, V. *Quart. Appl. Math.* **1996**, *54*, 225-247.
74. Pojman, J. A.; Willis, J. R.; Khan, A. M.; West, W. W. *J. Polym. Sci. Part A: Polym Chem.* **1996**, *34*, 991-995.
75. Pojman, J. A.; Khan, A. M.; Mathias, L. J. *Microgravity sci. technol.* **1997**, *X*, 36-40.

Chapter 2

Research Platforms That Reduce the Acceleration Experienced in the Experimental Frame of Reference: Achieving "Microgravity"

James Patton Downey

Microgravity Department, George C. Marshall Space Flight Center, SD48, National Aeronautics and Space Administration, Huntsville, AL 35812

Sedimentation, pressure gradients, i.e., pressure head, and buoyancy driven convection can be greatly decreased in experimental systems by accelerating the laboratory reference frame at a rate consistent with the acceleration due to gravity. This may be done in a number of ways, the best known of which is the use of orbiting spacecraft. Other techniques include the use of aircraft following an appropriate parabolic trajectory or drop towers. The result is an experimental condition in which fluids experience virtually no outside forces relative to the laboratory reference frame. Within this reference frame the effects of gravity have been reduced many orders of magnitude. Such conditions are appropriate for achieving conditions of diffusion-dominated heat and/or mass transfer, the study of phase transitions in the absence of pressure gradients, the study of solutal or thermal capillary convection, or containerless processes. Ways of achieving these conditions and complexities that arise in performing experiments using these research platforms are discussed.

Introduction

Acceleration due to earth's gravity affects solutions in several ways. For example, pressure gradients occur as a function of the height within a liquid column. Thus, gravity effects both equilibrium conditions and phase transitions as a result of hydrostatic pressure gradients. In addition, gravity affects the rate of heat and mass transfer in solution by buoyancy-driven convection. This can be particularly important in polymerization or polymer processing due to the high sensitivity of polymeric materials to processing conditions.

The term microgravity has been coined to describe an environment in which the effects of gravitational acceleration are greatly reduced. It may seem odd to talk in terms of reducing the effects of gravitational acceleration since gravitational attraction is a basic property of matter. However, the presence of gravity on in situ processing or measurements can be negated by achieving conditions in which the laboratory, or more specifically, the container of the experimental materials, moves at a rate consistent with the acceleration due to gravity. When this is achieved, the effects of gravity within the reference frame are negligible. This condition is difficult to achieve, but can be done. A short review of Newtonian physics provides an explanation on both how processes are effected by gravity and how so called microgravity conditions are achieved.

While all objects on the surface of earth experience the force of gravity, it goes without saying that we do not often find them accelerating at a rate of 9.8 m s^{-2}. The reason is that the objects are subject to other forces. A laboratory, its structures, and its containers are typically solid objects resting on a solid foundation. A property of solids that differentiates them from liquids is that solids resist being continually deformed by a shearing force. This property indicates that solids have a structure able to exert an opposing force that negates an externally applied force. Thus when a solid resting in a lab setting experiences the force of gravity, not only does it not accelerate at 9.8 m s^{-2}, it does not move at all.

Liquids are more complicated. Liquids can deform continuously when a force is applied, indicating that a liquid structure cannot completely negate an applied force. However, like solids it is not common to find conditions in which a liquid is accelerating at 9.8 m s^{-2}. A liquid's structure can oppose an applied force, but often this structure does not negate the applied force. Just how easily a liquid resists deformation due to an applied force such as gravity is related to its viscosity in the bulk of the fluid and its surface tension at interfacial surfaces of the liquid.

A liquid within a solid container is constrained by the solid structure of the container such that the liquid is confined. So under normal circumstances, the container exerts a force on the liquid opposing gravity. The sum of the forces on the fluid as a whole are zero. However, this does not necessarily mean that there is no acceleration within the liquid due to gravity.

All objects within the container experience identical gravitational acceleration due to gravity. However, the mass of two objects may be different. Thus, the gravitational force on the objects may be different. This imbalance of forces within a solution causes motion. Within a container, the more dense objects are subject to more force and settle to the bottom. The less dense objects are not subject to as much force and are displaced by the denser fluid and thus rise toward the top of the fluid. In the case of fluids, the difference in force results in the less dense fluid regions floating upward and the more dense fluid settling downward. This fluid movement is called buoyancy-driven convection.

Orbiting spacecraft provide an environment in which the laboratory reference frame, i.e., the center of gravity of the spacecraft, and all the equipment therein are experiencing virtually identical gravitational acceleration. There is no solid foundation underneath such a laboratory, so the laboratory accelerates according to the force of gravity as do the experimental fluids within the lab. Relative to the reference frame of the orbiting spacecraft, the forces on fluids within the spacecraft may be quite small. They are not zero however.

This is due to the fact that gravitational acceleration varies slightly with position within the spacecraft and that there is some force applied to the spacecraft due to atmospheric drag. When compared with a laboratory on the ground and averaged over time, the fluids in a spacecraft laboratory experience a reduction in acceleration relative to their laboratory reference frame by a factor of approximately 10^{-6}, hence the term microgravity. Achieving a low effective acceleration is not restricted to the use of spacecraft. Any laboratory that accelerates at a rate nearly consistent with the gravitational acceleration will result in a similar environment.

Experimental Platforms

There are several ways to achieve a reduction in the effective acceleration experienced by experiments. All have in common the establishment of conditions in which the laboratory is accelerating due to the force of gravity, but other forces on the lab are absent or insignificant.

Drop Towers

Drop towers are devices in which an experiment package is dropped down a long vertical path. These devices provide a few seconds of free fall in which the effective acceleration levels on the experiment package are quite small. To ensure that the experiment package drops at a rate nearly consistent with the acceleration due to gravity, the force of air resistance must be reduced, eliminated, or negated.

Some drop towers are evacuated of air. Others use a wake shield that travels just in front of the experiment package, producing a nearly drag free path for the experimental package. Another method that has been employed is to use air or gas jets attached to the package to counteract the force of air resistance. Finally, a mixture of these techniques has been used in which an experimental capsule is placed within a larger wake shield with the intervening volume evacuated. The wake shield is equipped with gas jets to produce force negating air resistance.

This latter combination is used in the largest drop tower in existence, one that can provide approximately ten seconds of low-g. It is located in Kamisunagawa, Japan, and run by the New Energy and Industrial Development Organization. (1) The package falls 490 meters, over a quarter of a mile, and is traveling at 980 m s^{-1} when braking commences. During the free fall, the interior package experiences an effective acceleration of about 10^{-5}g. The braking primarily uses air resistance as the tube narrows to a clearance of about 100 mm in the breaking section. The deceleration averages 3g with a maximum of 8 g. The maximum experiment mass for this drop tower is about 500 kg and must fit in a volume of 870x870x970mm.

There are other drop towers active in Bremen, Germany, Grenoble, France, and Cleveland, Ohio. The 5.18 second drop tower at NASA's Glenn Research Center (GRC) in Cleveland is known as the Zero Gravity Research Facility. (2) It can perform one to two drops per day. The experiment package drops down an evacuated shaft for 135 meters before breaking commences. The breaking occurs over a 10 meter distance with deceleration reaching 65g. This drop tower provides an effective acceleration level of about 10^{-5} g. The maximum mass of an experiment is 455kg and must fit inside a cylinder of 1meter diameter and 3.4 meter height.

The 2.2 Second Drop Tower at GRC *(2)* relies on a wake shield to reduce air resistance. It can perform up to 12 drops per day and can provide an effective acceleration of approximately 10^{-4} g where g is the acceleration due to earth's gravity at the earth's surface.

Parabolic Trajectory Airplane Flights

A plane can reduce the effective acceleration on experiments within by flying a trajectory in which the horizontal component of velocity is constant and the vertical component of velocity is consistent with acceleration due to the force of earth's gravity, i.e.,

$$dv_x/dt = 0 \; , \; dv_y/dt = g$$

where v_x and v_y are the horizontal and vertical components of velocity respectively, t is time, and g is the acceleration due to earth's gravity. Integrating these equations twice with respect to time reveals that the trajectory of the plane that satisfies these constraints is a parabola, i.e.

$$x = Ct \; , \; y = (1/2)gt^2$$

where the initial values of v_y, x, and y are assumed to be 0. NASA currently uses a KC135 jet often referred to as the "vomit comet" to fly parabolic trajectories providing an effective acceleration in the payload bay averaging about 10^{-2}g. (2) The plane can provide about 23 seconds of this acceleration level for each parabola. 27 seconds of low-g environment can be achieved on an individual parabola if specifically requested. The plane is occasionally used to provide "hypogravity" conditions. In this case, the plane flies a shallower parabola such that

$$y = (g-g_l)t^2 / 2$$

where $g-g_l$ is some value of acceleration typically consistent of some body of interest such as the surface of the moon or Mars. The plane typically flies about 40 parabolas per day at a cost of around $10,000 for a day of operation.

Spacecraft and Rockets

Spacecraft and rockets provide platforms with longer durations of low effective acceleration. Sounding rockets can provide 5 to 15 minutes of low g. The Space Shuttle can provide up to 17 days and the International Space Station (ISS) can provide months of low g. Average acceleration levels on spacecraft in low earth orbit can be on the order of 10^{-6} g. There are two sources of acceleration that account for this. One is atmospheric drag. The other acceleration effect is the so-called gravity gradient. These acceleration sources will be discussed in the next section.

Spacecraft are severely constrained experiment carriers. Flights are expensive, few, and far between. Safety considerations are paramount. In addition lead times in preparing for a flight are long. Resources required to perform experiments such as weight, volume, power, video observation time, commanding, crew time for operating an experiment, etc., are all in short supply. Costs of performing experiments on these platforms can be prohibitive. Sounding rockets costs are typically on the order of $100,000 to $1,000,000 per flight experiment. The use of the Space Shuttle or ISS is largely limited to parasitic use. When shuttle schedules or ISS utilization permits the performance of experiments, they are flown based on the availability of resources such as weight, power, volume, etc. These opportunities are made available only for selected experiments. Experiments in the NASA space program involving the shuttle, space station, KC135, drop towers, and related ground-based efforts are chosen based on a scientific peer review of a proposal. These are generally submitted in response to a NASA Research Announcement. These announcements are currently posted at http://peer1.idi.usra.edu.

Industry is allowed to purchase from NASA resources such as volume, power, etc. for proprietary research on spacecraft on a cost prorated basis or less if deemed in NASA's interest. Proof of concept flight experiments for industry may be selected based on the level of industrial commitment or participation by NASA's Space Product Development organization.

Acceleration Sources

The limited availability of low g platforms, the expense, and other constraints associated with their use are obvious from the previous discussions. A less obvious concern is the complexity of the heat and mass transfer associated with processes performed in these environments. There is a temptation to assume zero acceleration and further assume that all heat and mass transfer is diffusive. Often this is a bad assumption, particularly for the latter case. Frequently, the heat and mass transport occurring in a low-g facility is not easier, but harder to understand than that in the equivalent 1g experiment. This is because there are acceleration events that are transient, quasi-steady and oscillatory in nature. Each may be a relevant source of heat or mass transport. We will start the discussion with an overview of quasi-steady sources of acceleration.

Quasi-steady and Steady Acceleration

The most readily identifiable source of quasi-steady acceleration on a spacecraft is atmospheric drag. Even in low earth orbit there is sufficient atmosphere to slow the vehicle. Atmospheric drag varies according to the attitude and altitude of the vehicle. The value of atmospheric drag for the space shuttle tends to be on the order of tenths of microg. The acceleration due to drag is often referred to as quasi-steady since it varies slowly as the shuttle makes it orbit, a period of about ninety minutes. Drag varies as the shuttle goes from the sunlit side of earth to the dark. This variance can be as much as a factor of ten. Drag tends to be higher during times of high solar activity such as solar flares.

A major source of steady acceleration on experiments in spacecraft is often referred to as the gravity gradient. The acceleration of earth's gravity on an object at a location $x\mathbf{i}$, $y\mathbf{j}$, and $z\mathbf{k}$ from the center of the earth is

$$\mathbf{A} = - (Gm_e / \Sigma^{3/2}) (x\mathbf{i} + y\mathbf{j} + z\mathbf{k})$$

where $\mathbf{A}$ is the acceleration vector, G is the gravitational constant ($6.67 \cdot 10^{-11}$ m^{3}kg^{-1}s^{-2}), m_e is the mass of the earth ($5.9763 \cdot 10^{24}$ kg), Σ is equal to ($x^2 + y^2 + z^2$), and $\mathbf{i}$, $\mathbf{j}$, and $\mathbf{k}$ are unit vectors along perpendicular axes with origins at the center of the earth. $\mathbf{A}$ varies with the magnitude of x, y, and z according to

$$d\mathbf{A} / dx = - (Gm_e / \Sigma^{5/2}) \{(\Sigma-3x^2)\mathbf{i} - 3y^2\mathbf{j} - 3z^2\mathbf{k}\}$$
$$d\mathbf{A} / dy = - (Gm_e / \Sigma^{5/2}) \{-3x^2\mathbf{i} + (\Sigma-3y^2)\mathbf{j} - 3z^2\mathbf{k}\}$$
$$d\mathbf{A} / dz = - (Gm_e / \Sigma^{5/2}) \{-3x^2\mathbf{i} - 3y^2\mathbf{j} + (\Sigma-3z^2)\mathbf{k}\}$$

If we assume the spacecraft center of mass is at a location of $(x,y,z) = (R,0,0)$, these values become

$$d\mathbf{A} / dx = 2(Gm_e / R^3)\mathbf{i}$$
$$d\mathbf{A} / dy = - (Gm_e / R^3)\mathbf{j}$$
$$d\mathbf{A} / dz = - (Gm_e / R^3)\mathbf{k}$$

Since the gravitational acceleration varies slightly within the spacecraft, the spacecraft exerts a small force on its components to fix them in place relative to the spacecraft reference frame. Without this force, objects above or below the spacecraft's center of mass (relative to the earth's center) would drift further away from the center of mass. Objects in the horizontal plane perpendicular to the vector from the spacecraft's center of mass to the earth would drift towards the spacecraft's center of mass. At 300km above the earth's surface, the term (Gm_e/R^3) has a value of about $1.338 \cdot 10^{-6}$ s^{-2} or 0.1365 µg m^{-1} where µg is microg.

The above argument is appropriate for a case where the spacecraft is oriented toward some point in deep space. However, the shuttle, the Space Station, and most other spacecraft are usually oriented so that the same side faces earth for long periods. From the earth's frame of reference, the spacecraft appears to rotate once over the course of an orbit in order to keep the same face directed to earth. This has an effect on the quasi-steady acceleration environment. If at some moment the spacecraft is traveling in the **j** direction, then this rotation essentially negates the gravity gradient effect in the **j** axis.

The gravity gradient and atmospheric drag account for the micro-g rather than zero-g time-averaged acceleration environment found in a spacecraft. A spacecraft in a low earth orbit is typically about 300 kilometers above the earth's surface and about 6671 kilometers from the earth's center. It is interesting to note that r is only about 5% greater in low earth orbit than on the earth's surface. So the absolute value of **A** is only about 9% less in this orbit than on the ground.

Vibrational Sources of Acceleration

A more complex concern involves the effect of vibrations on experiments. In most ground experiments, an investigator can correctly assume that the greatest source of motion in otherwise quiescent experimental solutions arises from the constant acceleration of gravity and that the effects of vibrations are negligible. In low-g facilities, vibrations may be the most significant drivers of heat and mass transfer. There is a tendency to believe that since the vector average of a vibrational acceleration is zero, the effect on heat and mass transfer is zero. This would be wrong. It is true that liquids are somewhat impervious to high frequency vibrations. The time scale required for a liquid to deform under a vibrational force is L^2/ν, the characteristic length of the system squared divided by the kinematic viscosity. Hence on a short time scale little deformation occurs. However, lower frequency vibrations can mix solutions quite effectively. This is a complex problem involving the frequency of the vibration, the amplitude of the vibrational acceleration, the kinematic viscosity, and the geometry of the system.

At low frequencies a solution will have sufficient time to respond to the acceleration such that the maximum velocity of fluid flow attained will be approximately the same as for a steady acceleration. This flow will be proportional to the amplitude of the acceleration. At frequencies above some viscosity dependent cutoff value, the maximum fluid flow attained will still be proportional to the amplitude of the acceleration but inversely proportional to frequency of the acceleration. (3,4)

Accelerometers are used to characterize the vibrational levels associated with low-g experiment platforms. One useful analytical tool is to convert acceleration measurements via Fourier transformation into acceleration amplitude vs. frequency charts. Certain features tend to appear consistently in these measurements. Examples include vibrational modes associated with the flexing of spacecraft structural components, equipment operations, compressors, fans, etc.

The ISS was designed to minimize the effect of vibrations on experiments. Estimates of the effect of vibrations on the mixing of solutions were used to construct the acceleration requirements curve for the ISS. The ISS acceleration requirement provides an upper limit for the acceleration allowed as a function of frequency. However, typical amplitudes for acceleration on ISS are predicted to reach values greater than 10^{-2} g in certain frequency ranges, far above this requirements curve. Vibrational accelerations can be mitigated by isolation devices such as the Active Rack Isolation System (ARIS). ARIS is predicted to reduce the amplitudes of vibrations by one to two orders of magnitude for a broad frequency range.

A comparison of the ISS requirements curve, the non-vibration isolated ISS acceleration environment predictions, the ARIS vibration isolation acceleration predictions, and some sample shuttle data are shown in Figure 1. The shuttle data consists of several hours of acceleration data broken into 100 second increments that were each converted by Fourier transform from the time domain to the frequency domain. The bars indicate the 0, 25, 50, 75, and 100 percentile values for this shuttle data. That is 25% of the 100 second time periods sampled had acceleration levels for a given frequency fall between the first (lower) and the second (next lowest) bar on the graph.

Transient Accelerations

Spacecraft experience transient accelerations due to a number of events. These include dockings of spacecraft, attitude (orientation) adjustments, discharge of wastewater, the dissipation of spacecraft heat by boiling off of cooling water, etc. These events have a wide variety of duration, acceleration level, and frequency of occurrence. For example Space Shuttle waste water dumps occur about once a day with an acceleration of 10^{-6} g. Space Shuttle attitude adjustments typically occur every few minutes with a duration of about 0.25 seconds and an acceleration of about $4 \cdot 10^{-4}$ g. What these transient accelerations have in common with quasi-steady accelerations is that no vibration isolation can be used. They all involve a change in the spacecraft's

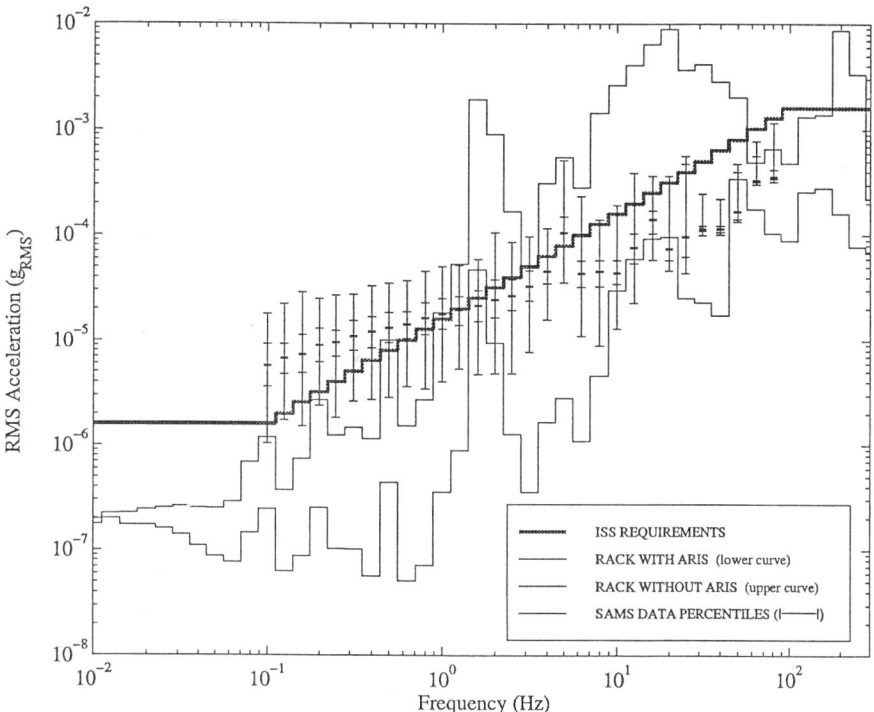

Figure 1. Comparison of Shuttle acceleration data with the predicted ISS acceleration environment. Representative shuttle data taken from the Space Acceleration Measurement System on the International Microgravity Laboratory 2 courtesy Glenn Research Center. ISS estimates courtesy of Johnson Space Center.

momentum. Hence, the average value of these accelerations is non-zero, unlike a vibrational acceleration. To a first approximation, the best way to judge the relative effects of these transients is to look at the resulting change of velocity, i.e., acceleration integrated over time.

Limitations of Acceleration Data

Acceleration data can be used to aid an investigator in understanding the acceleration environment of the experiment. However, this is limited to some degree by the fact that the accelerometer and the experiment are typically in different locations and experience different environments, particularly with regard to vibrations. Previous data gathered as part of earlier studies can occasionally be used to predict the acceleration environment for future experiments, but different experimental equipment, laboratory setup, and crew activities will result in changes that may not be predictable. To mitigate these limitations, optical tools are sometimes employed to aid in characterizing the heat and mass transport in low-g experiments. Still, investigators must often be content with performing experiments or creating materials under conditions of low levels of acceleration, the effects of which can only be roughly estimated.

Non-Dimensional Analysis

The low level of steady acceleration experienced on orbit may lead one to believe that heat and mass transport on orbit are dominated by diffusion. This is a dangerous assumption for an investigator to make. Diffusion is typically a slow, inefficient process. Common values for thermal diffusion coefficients in aqueous or organic solutions are on the order of 10^{-3} cm^2 s^{-1}. Values for thermal and mass diffusivity in aqueous solution are on the order of 10^{-3} and 10^{-5} cm^2 s^{-1} respectively. The associated diffusive velocities over a characteristic distance x are on the order of $v = D/x$, where v is velocity and D is the diffusion coefficient. Given the magnitude of D, v is generally small. Hence, even if the effects of acceleration are dropped six orders of magnitude, the reduction in buoyancy-driven convection may not be sufficient to reduce this source of heat and mass transfer to levels negligible in comparison with diffusive velocity. Similarly, the terminal velocity of settling particles may rival the diffusive velocity of interest. Given the complexities of the acceleration environment, an investigator must make use of non-dimensional analysis to determine if the necessary conditions for accomplishing the experiment are to be achieved.

A grossly simplified example may serve to illustrate how complicated the combined effects of diffusivity and convection can be. Consider the case of a fluid in a rectangular enclosure with a thermal gradient running perpendicular to the direction of gravitational acceleration. The vertical walls are conducting and of length H. The horizontal walls are insulating and of length L. For this geometry, the maximum velocity of the flow in the horizontal direction is known to be approximately (5,6)

$$v = g\beta H^3 \Delta T / \nu L$$

when

$$H/L > (g\beta H^3 \Delta T / \nu D_T)^{-1/4}$$

where g, β, ΔT, ν, and D_T are the acceleration due to gravity, the coefficient of thermal expansion, the temperature difference between the horizontal walls, kinematic viscosity, and thermal diffusivity respectively. The term within the parenthesis on the right is referred to as the Raleigh number. It is an estimate of the ratio of the heat transferred by convection and that transferred by diffusion.

For the sake of this argument assume that H is 1cm, L is 10 cm, and g is reduced to 10^{-3} cm s^{-2}, i.e., about a micro-g. Further assume that β is 10^{-3} C^{-1}, ΔT is 100C, ν is 1 cm^2 s^{-1}, and D_T is 10^{-3}cm^2 s^{-1}. Using these numbers we find v=10^{-5}cm s^{-1}. Compare this value with the diffusive velocity of D_T/L=10^{-4}cm s^{-1}. For these values, thermal transport is dominated by diffusion. However, convection is still an appreciable mechanism of thermal transport.

Now consider if there are concentration gradients in the system. Assume the diffusivity of mass is D_M = 10^{-5} cm^2 s^{-1}. The diffusive velocity of mass across this system is therefore approximated as D_M/L = 10^{-6} cm s^{-1}. Assuming that concentration gradients do not significantly effect the level of convection (an almost certainly wrong assumption unless the concentration gradients are small), we have a problem in which diffusion dominates thermal transport but convection dominates mass transport. In both cases the predominant source of transport is not expected to be the only relevant source of transport.

As in the above example, order of magnitude estimates can be made using the expected acceleration environment and then estimating the resultant heat and mass transfer. One useful non-dimensional number expresses the ratio of the buoyant force versus the viscous force in a system. This number is the Grashof number

$$Gr = g\Delta\rho L^3 / \rho\nu^2$$

Here $\Delta\rho$ and ρ are the density change over the length L and the density respectively. Depending of whether the buoyant force is caused by a thermal or concentration gradient, the ratio $\Delta\rho/\rho$ may be expressed as $\beta\Delta T$ or $\alpha\Delta C$ where α is the coefficient of thermal expansion and ΔC is the concentration change over the length L.

The surface tension analog of the Grashof number is the Marangoni number which is the ratio of the force of surface tension and viscous forces. The Marangoni number may be written as

$$Ma = \sigma\Delta TL(\partial\sigma/\partial T) / \rho v^2$$

where σ is the surface tension and $(\partial\sigma/\partial T)$ is the gradient of surface tension with respect to temperature.

A non-dimensional term useful for evaluating the effects of vibrations is the Strouhal number

$$St = \omega L^2 / v$$

where L is the length scale for the problem and is the ω frequency of the vibration in radians per second. When St is small, the vibrations should have little effect on the heat and mass transfer associated with the experiment. The reader may find the following references on the acceleration environment and its effects on heat and mass transfer useful. *(7,8)*

Discussion

The force of gravity has a profound influence on the processing of materials in solution. Thermal or concentration gradients in solution result in density gradients. This in turn leads to fluid flows effecting heat and mass transport as less dense fluid rises and more dense sinks within the solutions. These flows often convolute or mask other interesting phenomena such as diffusive processes or surface tension driven convection.

Gravity also results in pressure gradients in fluids. The pressure in a fluid is a function of depth. The greater the depth, the more mass overhead and hence the more weight of fluid pressing down from above. The resulting increase of pressure with depth is referred to as the pressure head or the hydrostatic pressure. Since pressure is an intrinsic thermodynamic quantity, hydrostatic pressure limits the uniformity of equilibrium conditions in a solution. For example, a phase transition in a solution varies as a function of pressure, hence depth within solution.

In addition, the effects of gravity typically require investigators to use solid containers when working with fluids. (The use of magnetic, electrostatic, or acoustic levitators is an exception to this.) The presence of a container can have major implications in terms of chemical compatibility concerns, e.g., semiconductor contaminants. Containers also influence the materials produced within, typically effecting properties such as grain structure, polycrystallinity, crystalline orientation, defect density, the degree of undercooling that can be achieved, etc.

Gravity is such a pervasive force that its ramifications on experiments are often not considered since the influence of gravity is perceived to be unavoidable. As a result, laboratory platforms that achieve an effective reduction of acceleration on experiments are often under appreciated. Based on the above arguments it is apparent that the reduction in the effects of acceleration on experiments can be used for studies of diffusive processes, surface-tension driven convection, containerless processing of materials, research on the scaling properties of phase transitions, etc. Space Shuttle experiments have been conducted involving each of the above topics.

The interested reader may wish to further study the diverse research involving these microgravity subject areas. The paper of Kamotani, et.al. *(9)* is recommended reading regarding microgravity experiments involving surface-tension driven convection. Likewise, a good source for reading on containerless processing experiments is *Solidification 1999 (10)* by Hoffmeister, et. al. A fine example of research involving scaling properties of fluids associated with phase transitions is the work of Chui, et.al. *(11)* on the Lambda Point of liquid Helium. The research of Zhu and Smith *(12)* involving diffusivity measurements is also recommended reading.

Summary

The influence of accelerations applied to experiments cannot be eliminated from the reference frame of the laboratory completely, but can be greatly reduced by the use of orbiting spacecraft, drop towers, or parabolic trajectory airplane flights. The relative merits of these devices are highly dependent on the experiments of interest. Previous experimenters have shown that a low acceleration environment can be useful in a number of areas. These include studies involving surface tension driven convection, containerless processing of materials, scaling properties associated with phase transitions, and achieving conditions of greatly reduced heat and mass transfer. In some circumstances, convective heat and mass transfer can be reduced such that diffusive transport dominates which allows accurate measurements of diffusion coefficients. Often, such experiments are either unachievable in a one g environment or of dubious accuracy due to the presence of convection, sedimentation, and/or pressure head. Due to the long lead times, the cost, and the complexity of conducting microgravity research in space, spaceflight research is advisable only for investigations in which there is no adequate alternative. Drop towers and parabolic airplane flights can often provide this alternative. However, when low acceleration is required for long duration, there is no alternative to the use of spacecraft.

References

1. See http://jamic.co.jp for a description of the NEDO drop tower.
2. See http://zeta.grc.nasa.gov/new/facility.htm for descriptions of the NASA's drop towers and the KC-135.
3. Griffin, P.R.; Motakef, S. *Appl. Microgravity Tech. II* **1989**, 3, 121.
4. Kondos, P.A.; Subramanian, R.S. *Microgravity Sci. Technol.* **1996**, 9, 143.
5. Cormack, D.E.; Leal, L.G.; Imberger, J. *J. Fluid Mech.* **1974**, 65, 209-229.
6. Began, A.B. *Convective Heat Transfer;* John Wiley and Sons: New York, NY, 1984; pp 159-18
7. Savino, R.; Monti, R. *Int. J. of Heat and Mass Transfer* **1999**, 42, 111.
8. Alexander, J.I.D. *Microgravity Science and Technology* **1990**, 3, 52.
9. Kamotani,Y.; Chang, A.; Ostrach, S. *J. Heat Transfer* **1996**, 118, 191.
10. *Solidification 1999*; Hofmeister, W.H.; Rogers, J.R.; Singh, N.B.; Marsh, S.P.; Vorhees, P.W., Eds.; The Minerals, Metals, & Materials Society: Warendale, PA, 1999; pp 3-106.
11. Chui, T.C.P.; Lipa, J.A.; Nissen, J.A.; Swanson, D.R. *Cryogenics* **1994**, 34, 341.
12. Zhu, X; Smith, R.W. *Advances in Space Research* **1998**, 22, 1253.

Sounding Rocket and Orbital Investigations

Chapter 3

The Role of Gravity in Sol–Gel Systems

Laurent Sibille[1] and David D. Smith[2]

[1]Universities Space Research Association, George C. Marshall Space Flight Center, SD48, National Aeronautics and Space Administration, Huntsville, AL 35812
[2]George C. Marshall Space Flight Center, SD48, National Aeronautics and Space Administration, Huntsville, AL 35812

> The evolution of sol-gel systems is regulated for the large part by the magnitudes of the interaction forces between sol particles. Since the viscosity of a gelling mixture increases during the process to approach infinity, the formation of entities such as aggregates or particles in the initial phase of gelation determines the physical properties of the final structure. In the midst of the many parameters affecting particle-particle interactions, gravity is one force, which although ubiquitous can be modulated independently. Depending on the magnitudes of the inter-particle interactions and the time scales involved, gravity may or may not play a significant role in the evolution of the final product by inducing convection and sedimentation for example. Nevertheless, gravity can be and has been used as an investigative tool in many important studies of sol-gel systems. While studies on the gravitational stability of large colloids have been published for some time, few groups however have used gravity as an investigative tool of phase diagrams for sol-gel systems. We review here the most significant studies in that field and report our most recent results on morphological changes in silica gels and Stöber particles obtained in microgravity.

The fast-growing interest for sol-gel science over the last twenty years is fueled by the extraordinary potential it offers for precise manufacturing control of a wide variety of modern materials. Over that span of time, we have seen vast improvements in products such as ceramics, glasses, optical devices, coatings and sensors when sol-gel technology has been successfully applied. While the list of successes increases and many research and development groups search for the next best formulation of the latest material, some have expressed concerns in recent conferences about our state of knowledge about the fundamental mechanisms of the solidification process in colloidal systems. Although particle interactions in colloidal systems have been well described by several efficient theories, large domains of phase diagrams remain unexplored or unexplained and the kinetics of gelling systems have only been described in a very partial manner. Our knowledge of these phase diagrams rests upon our understanding of the competition or cooperation between inter-particle forces and external forces such as gravity, which affect the system differently at the different stages of sol formation, aggregation and gelation. In that context, gravity must be considered as one of multiple variables such as pH, volume fractions of monomeric solutes, and solvent effects that create the wide range of conditions in colloidal systems.

Effect of Gravity on Sols

The initial phase of the evolution of a sol-gel system involves the formation of a sol, which is eventually driven to form a gel by various mechanisms. Simultaneous or successive effects of major forces determine the fate of the stable sol, even if stable only for a short time. The predominance of a specific force, whether internal or external to the system changes over the time of evolution of the sol toward a gel.

The DLVO model (named after its principal creators, Derjaguin, Landau, Verwey and Overbeek) is the most widely used to describe inter-particle surface force potential *(1,2)*. It assumes that the total inter-particle potential is the sum of an attractive van der Waals force and a repulsive double-layer force. The repulsive force due to the double-layer coulombic interaction between equal spheres separated by a distance D generates a positive potential energy V_R. If the radius r of the spheres is large compared to the double-layer thickness $1/\kappa$ ($\kappa r \gg 1$, with κ the Debye-Hückel parameter), V_R is described approximately by:

$$V_R = 2\pi\varepsilon r \psi_0^2 \ln[1 + \exp(-\kappa D_0)]$$

where ε is the permittivity of the dispersion medium, ψ_0 is the particle surface potential and D_0 the distance of closest approach *(3)*. These interactions between colloidal particles due to their double-layers can occur at distances of the order of 10/κ. Hamaker first derived the van der Waals potential between two isolated unequal spheres *(4)*. The range of the van der Waals force between atoms (or London dispersion force) is very limited due to its inverse dependence over the sixth power of the distance, thus virtually limiting its range to ~1nm. At first, such a force seems inadequate to counteract stronger coulombic forces and hold together large colloidal particles with rough surfaces, which rarely approach one another at these distances. However, the integration of the van der Waals forces between the atoms of the two approaching particles *(5)* yields the following potential energy of attraction $V_A = -Ar / 12D_0$ if the particles have an equal radius r. The rate of decay is therefore much slower and the resulting attractive force has an extended range of tens of nanometers, which rivals that of the coulombic forces. When both energy potentials are added as in the case of charged particles in a fluid, three possible behavior patterns are possible (Fig. 1): a) Formation of a (meta-) stable sol in which the potential energy barrier slows down the rate of coagulation of the particles; b) Formation of a marginally (un) stable sol and c) is undoubtedly unstable and undergoes rapid coagulation.

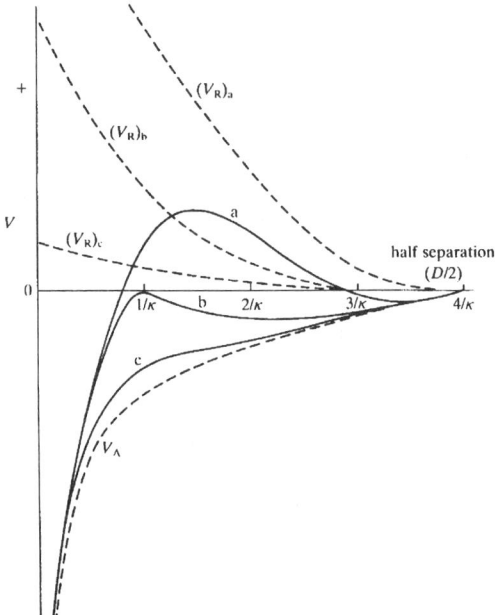

Figure 1. Total potential energy of interaction of a sol. (Reproduced with permission from reference 3. Copyright 1980 Plenum Press.)

It should be noted that the primary minimum imposed by V_A is not infinitely deep due to a very steep, short-range repulsion between the atoms on each surface. In addition, a term V_S, which describes the potential energy of repulsion due to adjacent solvent layers, becomes important at $D < 10$ nm. These forces become the 'choreographers' of the motions and behaviors of concentrated suspensions but they are part of a larger ensemble of forces and effects, which are very relevant for lower volume fractions of particles. At what stage does gravity play a role and what is that role among those interactions? From a theoretical point of view, the answer to that question is found by describing the motions of the particles in the surrounding fluid, i.e. one must solve the Stokes equations.

Mass transport: Convection and Diffusion

The treatment of fluxes of matter in fluids is of great importance in crystal growth studies where the transport of mass and heat to and across solid-liquid interfaces regulates growth rates and morphologies *(6)*. In the case of sol-gel systems, the early stages of molecular assembly in the system are affected by convective flows and Brownian diffusion as far as mass transport is concerned. The formalism describing these flows is derived from the principles of conservation of mass and momentum in the fluid. The conservation of mass in a non-isothermal multi-component fluid yields the *continuation equation* or *diffusion equation*, which expresses the rate of gain in mass density in terms of contributions from the convective flow and the diffusive flow. Similarly, the conservation of momentum in the system yields the *Navier-Stokes equation* describing the contribution of convection, pressure forces and viscous transfer and external forces on the rate of gain in momentum. The assumption of constant fluid density ρ for flow velocities that are very small compared to the speed of sound in that fluid yields simplified expressions of the two equations:

$$\nabla \cdot v = \frac{\partial v_x}{\partial x} + \frac{\partial v_y}{\partial y} + \frac{\partial v_z}{\partial z} = 0 \quad \text{(Continuity equation)} \tag{1}$$

$$\rho \frac{Dv}{Dt} = -\nabla p + \eta \nabla^2 v + F \quad \text{(Equation of motion)} \tag{2}$$

where v is the fluid velocity, p the pressure, F the sum of external forces and η the shear viscosity. The right side of equation (2) formulates the sum of the forces acting on the moving fluid mass in a unit volume. The negative of the pressure gradient is that force acting on the fluid when the pressure changes from one point to the next. The term $\eta \nabla^2 v$ (where ∇^2 denotes the Laplace operator, e.g., $\partial^2/\partial x^2 + \partial^2/\partial y^2 + \partial^2/\partial z^2$) describes the effect of viscous forces. On the other hand, the Lagrangian time derivative of the velocity vector includes an acceleration term and an inertia term ($Dv/Dt = \partial v/\partial t + (v \cdot \text{grad})v$). Sol-gel systems distinguish themselves from other systems such as crystallizing solutions by the fact that their viscosity may change with time as the solution evolves through aggregation, coagulation (or flocculation) and eventually gelation for some systems. Therefore, quantitative solutions of the equation of motion require the knowledge of the time evolution of the viscosity $\eta(t)$ through modeling or experimental data. Nevertheless, an attempt can be made to examine the effects of convective flows with respect to diffusive mass transport on evolving sol-gel systems by using Von Smoluchowski's theory for coagulation kinetics as a framework. The theory describing the kinetics of coagulation was developed by Von Smoluchowski who assumed that every collision between particles is effective and irreversible, i.e. dissociation can be ignored. The change in the population of single particles (singlets) A_1 from collisions with other singlets, doublets (two particles associated), and higher order aggregates is given by:

$$\frac{dA_1}{dt} = -2A_1 \sum_{i=1}^{\infty} k_{1i} A_i$$

where $k_{11} = 2\pi R_{11} D_{11} = k_s$ is the Smoluchowski rate constant, R_{11} is approximately twice the particle radius and defines a sphere of interaction where if the inter-particle distance d is less than R, particles will stick irreversibly, while D_{11} is the coefficient of relative diffusion between two (singlet) particles. Such coefficient takes into account all contributions to the motions of particles in the system, including that of convection. For example, the coefficient of relative diffusion and the rate constant k_S have larger values in cases where convective flows are significant in comparison with the Brownian motion of the particles. For spherical particles and approximately equal rate constants (i.e. $k_{ij} = k_s$) then we obtain for the total population of particles:

$$A_{total} = \sum_{i=1}^{\infty} A_i = \frac{A_0}{1 + k_S A_0 t}$$

where A_0 is the initial or steady-state singlet population. The change in the population of aggregates of type n then simplifies to:

$$A_n = \frac{A_0(k_S A_0 t)^{n-1}}{(1+k_S A_0 t)^{n+1}}$$

These results are only correct at the early stages of aggregation and are plotted in Figure 2. Note that we may define a coagulation time T_{ag} after which the total number of particles is reduced to ½ of the initial value.

$$T_{ag} \equiv \frac{1}{k_0 A_0}$$

At T_{ag} the number of aggregates of type n reduces to $A_0 / 2^{n+1}$. Note also that there is a maximum in the number of higher-order aggregates at $t = (n-1)T_{ag}/2$. The above model assumes that the coagulation is irreversible. However, if the potential energy barrier is not more that a few kT or if the minima are shallow then dissociation may occur. In the presence of sufficient repulsive forces only a fraction 1/W of particle collisions are effective and result in aggregation, where W is defined as the ratio of the diffusion flux in the limit of slow coagulation (repulsive forces considered) to the diffusion flux in the limit of rapid coagulation (repulsive forces not considered). If all other factors affecting the rate of collision remain unchanged, mass transport by convective flows also result in an increase in aggregation due in part to the larger frequency of collisions and in part to the increased kinetic energy of colliding particles.

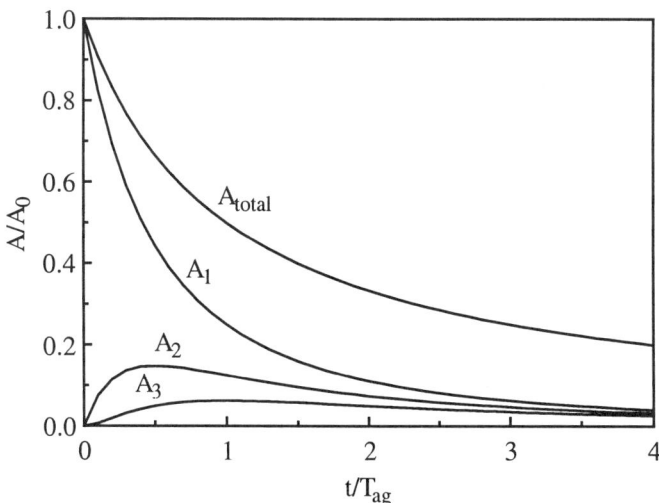

Figure 2. Von Smoluchowski Theory of Irreversible Coagulation

In consequence, the populations of higher order aggregates in Figure 2 will display maxima at shorter times. By comparison, the populations of lower-order aggregates will be relatively unchanged since convection affects most the relative diffusion of larger particles with respect to small ones.

Mass transport: Sedimentation and Brownian motion

The Navier-Stokes equations describe most of the commonly occurring flows from the large-scale oceanic motions down to the microscale of an experimental fluid cell. Solving the two equations to determine the velocity and pressure fields based on the knowledge of the long-range forces requires the knowledge of the boundary conditions. In our case, these conditions are imposed by the interface of the solid particles and the fluid, such as the 'no-slip' boundary condition. The problem of solving the Navier-Stokes equations for the flow around individual sedimenting particles is simplified by the fact that these flow fields have a very small length-scale. As a result, it is possible to neglect the inertia terms in the equation and arrive at the Stokes equation:

$$\eta \nabla^2 v = \nabla p - F \text{ and } \nabla \cdot v = 0$$

It is important to note that the Stokes equation is linear, unlike the original equation. Therefore the sum of two solutions is a solution itself, and we can treat the case of each external force acting on the system independently. Hence, one can study sedimentation without Brownian motion and vice versa. Solving the Stokes equations for a sedimenting spherical colloidal particle yields the following expression for the velocity of the fluid surrounding the particle:

$$v = \frac{2}{9} \frac{a^2}{h} \left(\frac{\rho_p}{\rho} - 1 \right) F$$

where a is the particle radius and ρ_p/ρ is the ratio of the density of the particle to the one of the fluid. Thus the sedimenting velocity of a particle will only be significant if its density is much larger than that of the fluid and/or when the particle reaches a certain size, which is the case for some micron-size colloidal particles. They will sediment even more quickly if the particles are aggregated to form floccules or flocs, a property often used in the early steps of washing a colloid. The Stokes equations have also been solved for ellipsoidal particles for which the sedimenting velocity depends on their orientation.

Coupling of Gravity with Colloidal Forces

Rigorous study of the kinetics of simultaneous Brownian and gravity-induced flocculation (i.e. sedimentation) requires solving the convective-diffusion equation, which describes the distribution of particles around a reference particle. Von Smoluchowski *(7)* worked the solution while ignoring interparticle forces and hydrodynamic interactions. Derjaguin and Muller *(8)* and Spielman independently *(9)* solved the equation in the steady state with no external forces (e.g. gravity, shearing motion, etc.). Internal forces were included in the solutions worked by Roebersen and Wiersma *(10)* and Feke and Prabhu *(11)*. The case of strong Brownian motion with weak shearing flows was studied by Van de Ven and Mason *(12)* and Batchelor *(13)* while Feke and Schowalter *(14)* have examined systems with strong shearing flows and weak Brownian motion. Melik and Fogler have provided a remarkable analysis of systems undergoing strong Brownian/weak gravity effects *(15)* and those under strong gravity/negligible Brownian effects *(16)*. They derive a single analytical expression for the coupled flocculation rates due to Brownian diffusion and the gravity field. The expression includes the total interparticle potential, which is formalized best by the DLVO theory in the case of colloidal systems. The equation allows the computation of the effect of gravity on the total flocculation rate Gr of a particle of radius a and a relative density $\Delta\rho$ compared to the solution: $Gr = 2\pi g \Delta\rho a^4 / 3kT$. Figure 3 represents the percentage increase $\Delta\%$ of the flocculation rate in function of the particle ratio $\lambda = a_1/a_2$, where a_1 is the radius of the reference particle and a_2, the radius of the particle in motion ($0<\lambda<1$). Remarkably, as an expression of the relative effect of gravity on the particles compared to Brownian forces, the flocculation rate increases with λ to reach a maximum, after which it decreases with increasing λ. However, they point out that while gravity can increase the probability density for a given inter-particle separation, which is the basis for the equation, gravity does not affect the actual flocculation process significantly. While this may seem counterintuitive, it stems from the fact that the major influence of gravity on flocculation is at large inter-particle distances only where particle trajectories follow only the undisturbed fluid streamlines. The authors reach the conclusion that while diffusion and gravitational creaming simultaneously bring the particles very close to each other, it is primarily diffusion and the van der Waals attractive force, which eventually overcome the repulsive forces to allow particle capture.

Gravitational Stability of Sols

The early experiments on the settling of dilute suspensions of fine particles in water performed by Jean Perrin *(17)* in 1909 quickly became controversial *(18)*. Perrin observed the sedimentation of colloids in a well thermostated cell of a thickness of 0.1mm. He showed that the suspension behaves like an ideal gas

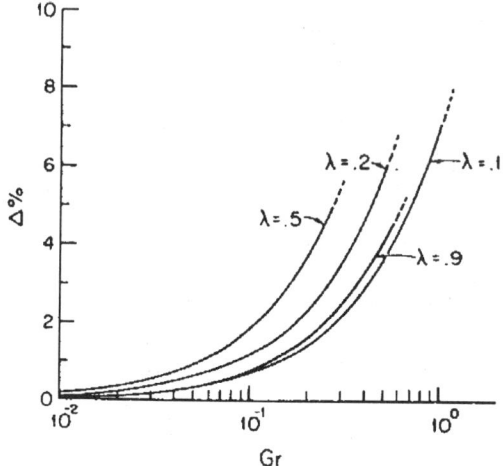

Figure 3. Percentage increase in the flocculation rate due to gravity for various particle size ratios. (Reproduced with permission from reference 15. Copyright 1984 Academic Press.)

of very large molecular weight and that the concentration of particles increases exponentially from top to bottom of the cell. Although confirmed by Westgren in 1915, they defied the common knowledge that colloidal suspensions appear uniform in concentration in large cells and special "long range colloidal forces" were once invoked to explain the discrepancy. In 1932, Usher and McDowell proved Perrin right by demonstrating that gravitational sedimentation occurs in colloidal suspensions over a wide range of concentrations and in cells of large (centimeter) dimensions when external influences such as vibrations and thermal fluctuations are eliminated. When returned to an uncontrolled environment, the suspension soon becomes homogeneous.

Met with great skepticism for more than twenty years, Perrin's experiments are no longer in doubt but the subject of gravitational stability of colloids is often treated as a curiosity or an afterthought in today's textbooks. Although the number of publications on gravity effects on sols is small compared to the whole of colloid literature, such studies shine the light on intrinsic properties of these solutions and allow quantitative measurements of fundamental interactions between particles.

Gelation and Gravity

A gel is widely defined as a giant molecule or cluster spanning the entire volume of the solution in which it originates. A wide variety of systems can generate a gel since the primary condition is that the starting monomer be able to make more than two bonds. According to Brinker and Scherer *(19)*, a gel is formed when the last bond that completes the giant molecule is formed and yields a substance containing a continuous solid phase enclosing a continuous liquid phase. Most gels can be categorized as being either particulate or polymeric in nature. The products obtained are very different from one another and the type of bonds that constitute their structure cannot define a gel. In fact, the bonds between particulate sols forming a network are often reversible (e.g. by shaking) due to attractive dispersion forces while polymeric gels are held by covalent bonds and gelatin gels by the restriction of motion of entangled chains. While the term 'semisolid' has been used, a more accurate definition is employed: a gel consists of continuous solid and liquid phases of colloidal dimensions. In such a system, the path one would take to cross either one phase perpendicular to its surface and re-emerge in the other phase is typically less than 1μm.

The Phenomenon of Gelation

We cannot review here the details of each model or approach used to describe gelling systems. Instead, our interest is in pointing out the assumptions made and how successful these approaches are in predicting the characteristics of the gels. In that respect, Zallen has provided a clear picture of the first two descriptions of the sol-gel transition: the classical theory and the percolation model *(20)*. The classical picture or 'mean field theory' developed by Flory and Stockmayer *(21)* is an oversimplification, which assumes that a chain is free to branch off, never to be limited by the growth of other branches in the same cluster. Moreover, the model assumes that bonds are formed only between polymers, never within them thus eliminating the possibility of closed loops. As a consequence, the theory provides adequate prediction of the critical degree of reaction (or minimum number of bonds formed per monomer to obtain gelation) only at high monomer concentrations when reactions between polymers are more likely than ones within themselves. Nevertheless, the Flory-Stockmayer theory of dendritic polymerization predicts well a sharp gel point and a very steep increase of the viscosity because of the divergence in the weight-average molecular weight at the approach of the sol-gel transition. However, the classical approach is also used to present a realistic picture of polymer growth in the gel because the assumption of indefinite growth leads to a polymer whose density increases proportionally to its radius.

The percolation model has been widely used to simulate and predict the evolution of a wide range of disordered systems from the spread of disease in a population to the glass transition *(22)*. Applied to the gelation phenomenon, it describes the formation of bonds at random along lattice lines. The bond percolation model on a three-dimensional lattice automatically includes the presence of rings or closed loops. Each lattice site is occupied by a monomer, which may form a covalent bond with any of its nearest neighbors. Like the classical mean-field theory, the percolation model predicts the sol-gel transition successfully but delivers different values for critical exponents essential to determine the increase of the gel fraction. Both the gel fraction increase and the divergence of the average molecular weight are much faster in the percolation model than in the classical view. De Gennes points out that such large discrepancies may be explained partly by the fact that many practical cases are intermediate between the two theories, ranging from systems of long linear chains to compact chains *(23)*. The percolation model remains a crude representation of the phenomenon of gelation with two major criticisms as noted by de Gennes: (i) in solution, the monomers are not distributed on a lattice but are disordered; (ii) the monomers are mixed in a solvent in most cases which is totally absent from the model. The ordered distribution of monomers is not as important for the kinetics of gelation as the ignorance of solvent interactions. As monomers link together covalently, they tend to form clusters in the solvent and thus to precipitate. Consequently, there is a competition between gelation and precipitation, which is important for many practical applications. Numerical data on this solvent effect shows that a trend toward segregation of the gelating species always develops *(24)*. However, the resulting phase diagram shows that one still obtains a well-defined sol-gel transition by suitable choice of the concentration of the gelling species. That insight into the kinetics of gelling systems points to the role played by gravity during the formation of critical size clusters as illustrated by a series of experimental studies.

Gravity Effects on Gelation

As pointed out before, the investigations of the effects of gravity on gelation are few and far between. Certainly, the notoriously arduous and costly access to the microgravity environment of orbiting spacecrafts is a serious limitation since it is one of the preferred approaches of study. Leontjev et al. reported the first microgravity gelation experiment using polyacrylamide (PAA) gels *(25)*. The photoinitiated process proved to be very sensitive to gravity because of the exothermic character of the condensation of monomers. Although the experimental conditions were not reported and such system cannot be considered a sol-gel in nature, the study showed a spectacular example of elimination of convective flows in microgravity resulting in the absence of "flow imprints" characteristic of side-irradiated PAA gels. When used for electrophoretic separation of parathyroid hormone from its precursors, the authors reported better resolution of the peaks by using the gels obtained in microgravity.

However, one can find even fewer experimental data on the behavior of gelling solutions in hypergravity, for which these obstacles don't exist. Hecht and Geissler *(26)* reported an interesting study of gelation of poly(acrylamide/bisacrylamide) under external forces, namely irradiation by Xray and centrifugation. Using a Schlieren optical detector to measure the change of refractive index along the length of the cell during centrifugation between 7000 and 20000 rpm, they observed a multiple-gelation behavior where successive layers of gel form at the gelation interface. Because of crosslinking, the layered structure was preserved once the gels were retrieved as opposed to mixtures of sols in which compression effects were relieved at the end of the experiment. Since the gel was formed in a compressed state, the only change of refractive index was observed at the sol-gel interface when new gel formed. In the case of such an experiment, the sol solution above the compressed gel acts as a reservoir of fresh monomers and oligomers diffusing toward the interface. Under normal or sub-normal gravity levels, the gel phase occupies the whole cell volume. The discrete occurrence of the gelation at the interface may be attributed to the diffusion delay of the monomers through the exiting gel layers and to the nonlinear nature of the gelation process. This behavior is closely related to the oscillatory phenomenon of Liesegang rings, which form under similar conditions *(27)*.

In our laboratory, we have recently conducted gelation studies of silica nanoparticles in microgravity during the STS-95 space shuttle mission *(28)*. Stable silica nanoparticle dispersions may be formed either by polymerization of silicic acids in an aqueous system or through hydrolysis and condensation of silicon alkoxides (the sol-gel or Stöber route). Comparison of small-angle x-ray scattering (SAXS) measurements of Ludox, a commercial aqueous silicate with acid- and base-catalyzed alkoxides shows that only aqueous silicate sols are uniform, whereas alkoxides generate fractal particles. As Brinker and Scherer point out *(29)*, these results illustrate that sols derived from aqueous silicates are fully hydrolyzed and grow by classical monomer addition resulting in uniform polymeric particles, whereas sols derived from silicon alkoxides grow through cluster aggregation and retain a fractal inner morphology even while the particles coarsen through surface tension reorganization. Microgravity allows diffusion-limited conditions to persist in recipes that typically are reaction-limited, essentially expanding the parameter space under which diffusion-limited conditions prevail, and providing a snapshot of the aggregation process that would not normally be accessible. We obtain good quality silica particles in the range 100 - 700 nm using the Stöber method in four different formulations on the ground. However, at certain precursor concentration ratios the particles are either polydisperse, bimodal, rough, or partially aggregated. Hence the recipes were carefully chosen to examine these 'failure conditions', essentially spanning a large portion of the parameter space over which Stöber particles may be produced. The tetraethylorthosilicate (TEOS) was freshly distilled and the samples were sealed in quartz glass under nitrogen to prevent hydrolysis prior to the experiment. The first recipe (R1) was a control sample chosen to produce the best possible particles in terms of monodispersity and sphericity. The second recipe (R2) was chosen to produce the smallest Stöber particles, which tend to be rough, irregular, and less monodisperse. Recipe R3 was chosen to produce a bimodal size distribution, while recipe R4 was chosen to produce large irregular (non-spherical) particles. The composition and molar ratios of each recipe is shown in Table I.

Table I. Molar Composition of Silica Sol-Gel Solutions

Recipe	TEOS (10^{-6} mole)	Ethanol (10^{-5} mole)	EtOH/ TEOS	H_2O (10^{-5} mole)	H_2O/ TEOS	NH_4OH (10^{-5} mole)	NH_4OH /TEOS
R1	0.63	7.21	114.2	0.0	0.0	1.67	26.4
R2	0.69	7.86	113.8	0.0	0.0	0.62	9.0
R3	2.01	6.52	32.5	3.20	16.0	0.43	2.1
R4	1.51	6.13	40.6	2.33	15.4	1.70	11.2

Identical solutions yielded structures of dramatically different morphologies when processed in 1g and microgravity as shown in Transmission Electron Microscope images for recipes R1 and R2 (Fig. 4).

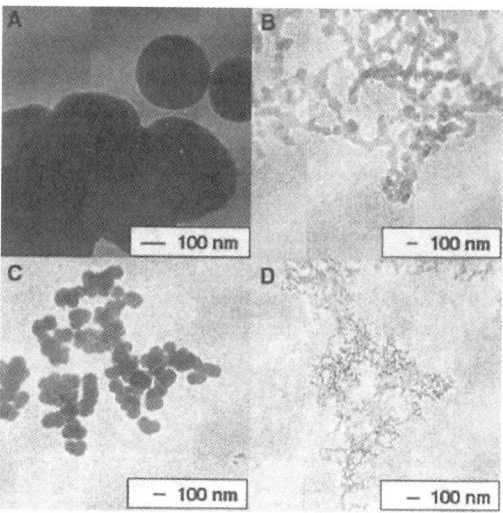

Figure 4. Silica sol-gel structures in 1g and microgravity(μg). Recipe R1 in 1g (A) and in μg (B). Recipe R2 in 1g (C) and in μg (D)

Visual inspection revealed that each of the space-grown samples had formed marginally coherent low-density gels, forming a continuous network percolated by the solvent. Ground-grown samples on the other hand formed large spherical particles. Whereas growth in unit gravity produces Stöber particles, growth in microgravity favors loose gel structures. In microgravity however, all recipes produced Stöber particles embedded in the tenuous gels, albeit in much fewer were found in R1 and R2, which did not contain additional water (the ammonium hydroxide reagent also contains water). All space-grown samples did form gels, and these gels had a common form and scale, which was nearly recipe independent. The particles making up the backbone of the gel were elongated with diameters of approximately 10 nm and lengths of about 50 nm. These gels are similar to the structures observed by Yoshida who added a calcium salt and a sodium hydroxide solution to a polymerizing silicic acid sol and heated the mixture in an autoclave to effect non-uniform particle growth *(30)*. For the recipes containing added water, R3 and R4, the particles making up the gel backbone were slightly wider and less elongated and surrounded more large spherical particles (Fig. 5). These spheres tended to be smaller (in some cases only about half as large) and have much larger size distributions than Stöber particles formed under identical conditions on the ground.

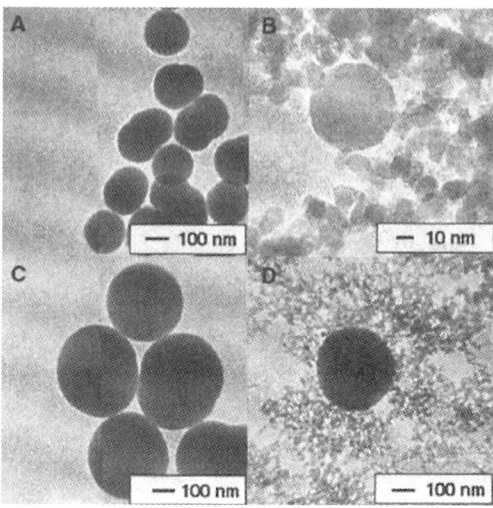

Figure 5. Silica sol-gel structures in 1g and microgravity(μg). Recipe R3 in 1g (A) and in μg (B). Recipe R4 in 1g (C) and in μg (D)

As shown by Bogush and Zukoski *(31)*, the coexistence of large monodisperse spheres (50 - 250 nm) with smaller (~ 10 nm) aggregating primary particles implies that the growth of silica Stöber particles does not occur by the classical nucleation and growth model, where a fixed number of particles are produced in a single event. Rather Bogush and Zukoski deduced that nucleation of particles proceeds continuously throughout most of the reaction period. The smaller primary particles form by the classical monomer addition growth mechanism and then aggregate because of their small size, until they become colloidally stable. Bogush and Zukoski propose that the resulting stable aggregates are the building blocks for the formation of Stöber particles, collecting smaller aggregates and newly formed particles as they are transported through the solution. Therefore, in this view, reaction-limited conditions must persist to maintain smooth spherical particles. The final structure then coarsens through surface tension reorganization to form the resulting Stöber particle.

In contradiction with Bogush and Zukoski, van Blaaderen and Vrij *(32)* have argued that if growth continued to occur through aggregation of subparticles, smooth spherical particles cannot result. In their view Stöber particles initially grow by aggregation of subparticles but monomer addition later fills in the nonuniformities, resulting in a smooth particle. The irregular shape of smaller Stöber particles is a remnant of the aggregation mechanism not yet enveloped by the subsequent monomer growth. The size difference between the subparticles and the resulting Stöber particles certainly supports this view, since this difference is likely too small even for reaction-limited conditions to yield a smooth surface. Furthermore, although only a few Stöber particles formed in microgravity, those that did form were smooth. Hence, the fact that smooth particles are obtained even in the absence of reaction-limited conditions further supports this view. The importance of the aggregation of unstable clusters and subparticles to the formation and growth of silica Stöber particles makes the effect of microgravity on Stöber particle growth profound. Rather than simply retarding structure growth (in this case a silica sol) as would be expected for a singular growth mechanism, a pathway to an entirely different structure becomes available. Microgravity favors diffusion-limited conditions, which slows the formation of stable particle-forming aggregates. Monomers are consumed more by unstable subparticles and aggregates than by Stöber particles. Eventually cluster-cluster aggregation is the only remaining growth mechanism, which yields more extended structures leading to a decrease in the fractal dimension and ultimately to gel formation. We observed gels in microgravity at TEOS concentrations as low as 2.8 vol.%. These results suggest that microgravity favors the formation of more rarefied sol-gel structures, providing a bias towards diffusion-limited cluster-cluster aggregation. Indeed the softest of matter may only be achievable in microgravity, where entropic reductions and perturbations to structure formation are minimized.

Using the same access to microgravity on STS-95, Segrè et al. *(33)* have studied the aggregation behavior of attractive polymethyl methacrylate (PMMA) spheres in near isopycnic conditions over a wide range of volume fractions and attractive energies. They have found that the solidification process in moderately attractive colloidal suspensions displays many of the features previously associated with the colloidal hard sphere glass transition. In such a system, particle bonding is reversible and the evolution of the structures formed, fluid or solid clusters, is very sensitive to the bonding strength. From comparative studies on the ground and in microgravity, they observe that the results are not significantly affected by gravity if the density match $\Delta\rho/\rho$ is less than 0.01.

Allain et al. *(34)* reported one of the first studies of the effect of gravity in polymer gels near the sol-gel transition. Inducing the copolymerization of acrylamide and bisacrylamide, they have observed the segregation of gel phases proposed by de Gennes. Moreover, they report the gravitational collapse of the gel phase under its own weight and indicate that the sedimentation was observed at the end of the chemical process. As part of their investigation, they found that the volume fraction to obtain gelation under gravity is much higher than one in the absence of gravity, which was indirectly confirmed in the gels of silica nanoparticles we obtained only in microgravity.

In the field of biopolymers, Roedersheimer et al have reported strong variations of morphology for both fibrin *(35)* and collagen *(36)* gels produced in microgravity flight. In a purified form, collagen and fibrin can be grown into gel-like matrices of interconnecting fibers. The flight samples had a more uniform morphology in the micron as well as the millimeter length scale as the ground control samples. Specifically, the authors found that in the case of collagen both the direction of diffusion of buffer ions required to create the conditions for self-assembly and gravity were strong factors in controlling the porosity and uniformity of the matrix. The porosity of both collagen (1-10 μm pores) and fibrin (20-60 μm pores) gels is in the range making them most susceptible to sedimentation effects. By contrast, the gels formed on the ground exhibit large voids as large as a millimeter! It is noteworthy to mention that the initial microgravity experiments performed by the same authors with collagen led them to conclude that there was no discerning effect of gravity on the porosity. In a later publication, they indicated that the use of experimental cells that were too small was probably responsible for masking the gravitational effect. One must realize that the absence of gravity does not change the dimensions of the building units, i.e. the fibers: instead, it affects the assembly of these fibers into the final gel product. This important distinction is generally true for macromolecular assemblies and gelling systems.

Conclusion

In spite of the fragmented picture painted by the few studies on the effect of gravity on the sol-gel transition, the data obtained thus far reveals that the gravitational field plays a ubiquitous role throughout the evolution of sol-gel systems. However, the effects of gravity may be predominant or weak depending on solution parameters such as particle size ratios and charges, relative density and viscosity. Convection will increase the rates of collision of small particles while sedimentation will become a factor for larger particles with significant relative densities. Gravity forces may not play a significant role near the gel point in one system while affecting the weakly connected network of another. At times, these effects can be adequately handled by density matching techniques or used to our advantages, but the broad questions still formulated at the conclusion of the most recent studies are indicative of our lack of understanding of this basic effect with wide relevance for many modern nanostructured materials.

Acknowledgements

We thank Raymond J. Cronise for his relentless efforts to assure the success of the STS95 mission, Lanee A. Snow for her dedication in preparing the microgravity experiments and astronaut John H. Glenn for conducting these experiments in orbit on STS95. We also thank Arlon J. Hunt and Michael R. Ayers for sharing their knowledge of sol-gel systems with us and Naomi J. Halas for her collaboration on the formation of Stöber particles. Financial support was provided by the Space Product Development office at MSFC and by NASA under NRA-96-HEDS-02.

References

1. Verwey, E. J. W.; Overbeek, J. Th. G. *Theory of the Stability of Lyophobic Colloids;* Elsevier: New York, NY, 1948.
2. Derjaguin, B.; Landau, L. *Acta Physicochim.* **1941**, 14, 633.
3. Hunter, R. J. In *Comprehensive Treatise of Electrochemistry;* Bockris, J. O'M.; Conway, B. E.; Yeager, Y., Eds.; Plenum Press: New York, NY, 1980; Vol. 1, pp 397-437.
4. Hamaker H. C. *Physica* **1937**, 4, 1058.
5. Hunter, R. J. *Foundations of Colloid Science;* Oxford University Press: New York, NY, 1992, Vol. 1, p 99.
6. Rosenberger, F. In *Fundamentals of Crystal Growth I: Macroscopic Equilibrium and Transport Concepts;* Cardona, M.; Fulde, P.; Queisser H. -J., Eds.; Solid State Sciences Series 5; Springer-Verlag: Berlin, Germany, 1979; Vol. 1, pp 215-295.
7. Von Smoluchowski, M. *Z. Phys. Chem.* **1917**, 92, 129.
8. Derjaguin, B. V.; Muller, V. M. *Dokl. Akad. Nauk. S.S.S.R. (English translation)* **1967**, 176, 738.
9. Spielman, L. A. *J. Colloid Interface Sci.* **1970**, 33, 562.
10. Roebersen, G. J.; Wiersema, P. H. *J. Colloid Interface Sci.* **1974**, 49, 98.
11. Feke, D. L.; Prabhu, N. *AIChE Annual Meeting Proc.* Nov. 14-19, 1982.
12. Van de Ven, T. G. M.; Mason, S. G. *Colloid Polym. Sci.* **1977**, 255, 794.
13. Batchelor, G. K. *J. Fluid Mech.* **1977**, 83, 97.
14. Feke, D. L.; Schowalter, W. R. *J. Fluid Mech.* **1983**, 133, 17.
15. Melik, D. H.; Fogler, H. S. *J. Colloid Interface Sci.* **1984**, 101, 84-97.
16. Melik, D. H.; Fogler, H. S. *J. Colloid Interface Sci.* **1984**, 101, 72-83.
17. Perrin, J.B. *Ann. Chim. Phys.* **1909**, 18, 5.
18. Mysels, K. J. *Langmuir* **1992**, 8, 3191-3194.
19. Brinker, C. J.; Scherer, G. W. *Sol-Gel Science: The Physics and Chemistry of Sol-Gel Processing*; Academic Press: San Diego, CA, 1990; p 8.

20. Zallen, R. *The Physics of Amorphous Solids*; John Wiley & Sons: New York, NY, 1983; pp 135-204.
21. Flory, P. J. *Principles of Polymer Chemistry;* Cornell University Press, Ithaca, NY, 1953; chapter IX.
22. Meakin, P. *Fractals, scaling and growth far from equilibrium;* Cambridge Nonlinear Science Series 5; Cambridge Univ. Press: Cambridge, UK, 1998.
23. De Gennes, P-G. *Scaling Concepts in Polymer Physics;* Cornell University Press, Ithaca, NY, 1979; 137-160.
24. Lubensky, T. C.; Isaacson, J. *Phys. Rev. Lett.* **1978**, 41, 829.
25. Leontjev, V. B.; Abdurakhmanov, S. D.; Levkovich, M. G. *AIAA/IKI Microgravity Science Symp. Proc.*, Moscow, May 13-17, 1991, 273-277.
26. Hecht, A. M.; Geissler, E. *Macromolecules* **1987**, 20, 2485-2490.
27. Fenney, P. J.; Gilbert, R. G.; *J. Colloid Interface Sci.* **1985**, 107, p 159.
28. Smith, D. D.; Sibille, L.; Cronise, R. J.; Hunt, A. J.; Oldenburg, S. J.; Wolfe, D.; Halas, N. J. *Langmuir* **2000** (accepted).
29. Brinker, C. J.; Scherer, G. W. *Sol-Gel Science: The Physics and Chemistry of Sol-Gel Processing*; Academic Press: San Diego, CA, 1990; pp 97-228.
30. Yoshida A. In *The Colloid Chemistry of Silica;* Bergna, H. E., Ed.; American Chemical Society: Washington, 1994; p 51.
31. Bogush G. H.; Zukoski, C. F. In *Ultrastructure Processing of Advanced Ceramics*; Mackenzie, J. D.; Ulrich, D. R., Eds.; Wiley: New York, NY, 1988; p 477.
32. Van Blaaderen, A.; Vrij, A. In *The Colloid Chemistry of Silica;* Bergna, H. E., Ed.; American Chemical Society: Washington, 1994; p 83.
33. Segre, P. N.; Prasad, V.; Schofield, A. B.; Weitz, D. A. *Phys. Rev. Lett.* **2001**, *to be published*.
34. Allain C.; Amiel, C. *Phys. Rev. A* **1990**, 42, 2, 843-848.
35. Nunes, C. R.; Roedersheimer, M. T.; Simske, S. J.; Luttges, M. W. *Microgravity Sci. Technol.* **1995**, VIII/2, 125-130.
36. Roedersheimer, M. T.; Bateman, T. A.; Simske, S. J. *J. Biomed. Mat. Res.* **1997**, 37, 2, 276-281.

Chapter 4

Growth Kinetics of Polydiacetylene Films Prepared in Microgravity

William E. Carswell[1], Mark S. Paley[2], Donald O. Frazier[3], and Robert J. Naumann[1]

[1]Alliance for Microgravity Materials Science and Applications, The University of Alabama in Huntsville, 301 Sparkman Drive, RI-M34, Huntsville, AL 35899
[2]Universities Space Research Association, George C. Marshall Space Flight Center, SD48, National Aeronautics and Space Administration, Huntsville, AL 35812
[3]George C. Marshall Space Flight Center, SD48, National Aeronautics and Space Administration, Huntsville, AL 35812

A diffusive/kinetic rate equation was developed for the growth of polydiacetylene films from solution and compared with a microgravity experiment. The model takes into account both the kinetics of thin film growth and the diffusive transport limitations inherent to microgravity. In order to apply this model, measurements of the density and the ultraviolet extinction coefficient of the films, as well as of the diffusion coefficient of the monomer/solvent system, were made. The thin films grown in microgravity were predicted by the model to grow to a thickness of 0.148 μm, versus 0.150 μm for the ground control films. The flight films grew to 0.102 μm

Introduction

This work is an extension of the work carried out by Paley et. al (*1, 2*) involving the novel, patented (*3*) technique for preparing films of polydiacetylenemethylnitroaniline (PDAMNA) from solution in 1,2-dichloroethane (DCE). The films are of interest for their significant nonlinear optical properties. The diacetylenemethylnitroaniline (DAMNA) monomer is shown in Figure 1.

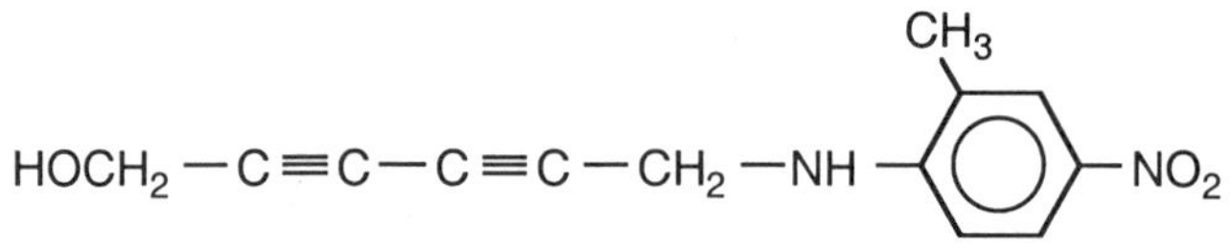

$$HOCH_2-C\equiv C-C\equiv C-CH_2-NH-\underset{}{\underset{}{\bigcirc}}-NO_2$$
with CH_3 substituent on the ring

Figure 1. DAMNA Monomer.

The films are grown through a straightforward ultraviolet polymerization process, shown schematically in Figure 2. A longwave fluorescent bulb, putting out a band around 365 nm, provides the UV light, approximately 5.1 mW/cm^2, for the microgravity experiment. The quartz substrate caps off one end of a cylindrical aluminum chamber containing the DAMNA/DCE solution from which the films are grown. When the chamber is irradiated with the UV through the quartz window the DAMNA begins to polymerize. The polymer that forms near the quartz substrate sticks to the substrate, forming an amorphous film. Other polymer is formed farther from the substrate and is unable to be incorporated into the growing film. In that case the polymer is left suspended in the bulk DCE/DAMNA solution. This polymer remaining in solution is relatively insoluble in DCE and quickly aggregates into larger particles, approximately 100-200 nm in diameter (*4*). Aggregation has the effect of reducing the number concentration of particles available for film growth.

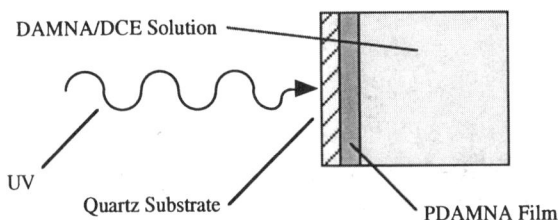

Figure 2. Thin Film Growth Schematic.
(Reproduced with permission from reference 4. Copyright 2000.)

Microgravity Justification

Work has been done in the area of numerical convective modeling for the DAMNA/DCE system in the cell being used for these experiments (5). This analysis shows that convective flows in the laboratory reach a maximum of 1.4 cm/sec with a UV intensity of 1.0mW/ cm^2 and a monomer concentration of 2.5 mg/ml. The microgravity flight and associated ground-control experiments used the same 2.5 mg/ml concentration, but the UV intensity for those experiments was 5.1 mW/cm^2. Since the flow velocity in the horizontal heated wall system scales roughly linearly with temperature gradient, and the heated wall temperature scales roughly linearly with UV light intensity, the maximum flow rate in the ground-based experiments is expected to be about five times the value obtained in the published model, or about 7 cm/sec. In order to achieve diffusion-controlled transport conditions, the convective flows need to be at least one order of magnitude lower than the diffusive transport in solution. Referring forward to laboratory results, the diffusivity of DAMNA in DCE is about $2 \cdot 10^{-5}$ cm^2/sec. This translates into a mean displacement of $(2 \cdot 2 \cdot 10^{-5}$ cm$^2)^{0.5}$ in one second, or a displacement velocity of roughly $9 \cdot 10^{-3}$ cm/sec. Therefore the convective flow needs to be a maximum of 10^{-3} cm/sec in order for solutal distribution to be governed by diffusion rather than convection. Since convection scales approximately linearly with gravity, there must be a reduction of four or five orders of magnitude in g in order for convection to be negligible. These conditions are readily achievable in the space shuttle, and therefore the flight experiments can be treated as convection-free and the model needs no adjustments for convection.

Performing the experiment in the classical thermally stable top-heated configuration was considered for this experiment, but discarded as ineffective for two reasons: non-uniform heating and sedimentation. In the ideal world an experiment is convectively stable from a thermal point of view when performed in such a way that the system is hotter on the top, cooler on the bottom, and contains no radial thermal gradients. This experiment is heated in the area of the film, and so it at first appears that it can be performed in a thermally stable configuration by irradiating it with the window facing up with respect to gravity. However, heat is supplied not by a geometrically uniform heat source, but by absorption of the incoming UV light. Hence the walls of the system are never heated, except by conduction from the fluid. This causes the walls to be at a lower temperature than the center of the film/fluid interface, setting up radial thermal gradients, which in turn drive convective flows. Sedimentation also plays a role that cannot be eliminated by performing the experiment in a top-heated orientation. The density of the polymer aggregating in solution is greater than the density of the solution, causing sedimentation and affecting the results.

Regarding the possibility of modifying the experiment chamber according to a scaling analysis to provide convection free transport on Earth, the complex nature of this experiment makes this approach unsuitable. The problem lies in the fact that this is not a heated wall fluid flow problem with a superimposed concentration depletion profile problem. This system is much more complicated.

The thermal convection scaling analysis for this system is based on the thermal Grashof number (6), shown in Equation 1,

$$Gr_t = \frac{g \cdot K \cdot L^3}{v^2} \cdot \Delta T, \qquad (1)$$

where g is the acceleration due to gravity, K is the thermal expansion coefficient of the fluid, L is the characteristic dimension of the convection chamber, in this case the depth of the PTFG cell, v is the viscosity of the fluid, and ΔT the difference in temperature between the heated wall the parallel wall bounding the back side of the convection chamber. Notice the implicit assumption of a heated wall by the ΔT term.

This brings up the first complicating issue, the fact that this is not a heated wall convection problem in the first place. This system is heated by incoming UV light that is absorbed by both the growing film and the monomer in the bulk solution. During the initial stages of growth the film is nonexistent, or nearly so, and the UV light is absorbed almost totally by the DAMNA monomer in solution. With a concentration of 2.51 mg/ml, the effective absorption coefficient, ε, of the solution is 157 cm^{-1}. Using the exponential intensity attenuation (absorption) relationship, $I = I_o \cdot e^{-\varepsilon \cdot l}$, where I is the UV intensity in solution as a function of penetration depth l, with an incident intensity I_o equal to 5.1 mW/cm^2, the intensity distribution in solution is shown in Figure 3.

When the heated region of the solution is narrow as compared to the dimensions of the convection cell, the heated wall assumption is acceptable. But when the heated region of the solution is large as compared to the dimensions of the cell, as might be the case if this experiment were scaled to reduce convection, the scaling analysis becomes invalid because the heated wall assumption becomes invalid.

It has already been established that convection, and hence the Grashof number, needs to be reduced by four to five orders of magnitude in order for diffusive mass transport to dominate in the system. This results, according to Equation 1, in a reduction of the characteristic length dimension of the cell from 1cm to 0.02-0.05 cm. Looking at Figure 3, clearly the UV light is absorbed far from the "heated wall," and the heated wall approximation for the system is no longer valid. Thus, by scaling the system, it has been transformed from the system under study into an incomparable system.

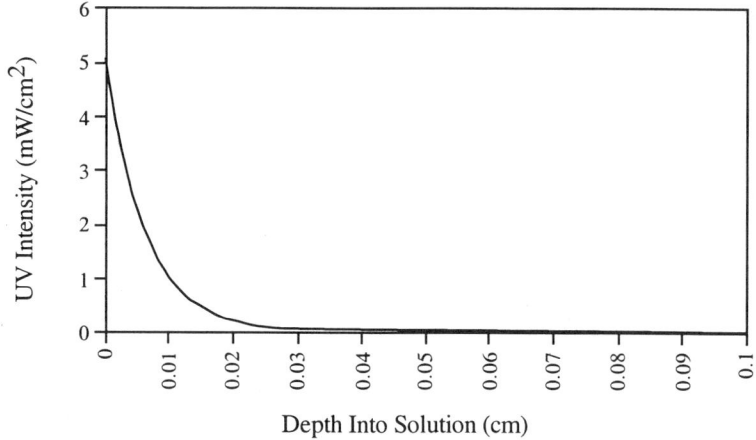

Figure 3. UV intensity as a function of penetration depth in a PTFG cell.
(Reproduced with permission from reference 4. Copyright 2000.)

The second problem with a scaled experiment arises from the presence of the aggregation process. Aggregation takes place everywhere polymer is present in the bulk solution. The bulk solution polymerization process, however, takes place only immediately in front of the quartz window, where the UV light is present in meaningful amounts. Under well-stirred conditions this polymer is evenly distributed throughout the system by convection and plays no real role in the reduction of the bulk concentration. In a convection-free environment, however, such is not the case. Granted, this is true whether convection is eliminated by reducing gravity or by a reduction in the characteristic dimension of the system due to a scaling analysis. The difference between the two approaches, however, lies in the ability of the aggregates to migrate away from the growing film in microgravity. The aggregate concentration, in a convection-free environment, is at a peak very near the interface, as shown by Figure 3. The gradient from the peak toward the growing film, which causes diffusion of aggregate toward the film, is caused by the incorporation of aggregate into the film. The gradient away from the film and into the bulk solution is governed by the ability of the aggregate to diffuse into the bulk solution. If the chamber dimension is reduced to the same order as the size of the aggregate concentration peak, the diffusion of aggregate away from the growing film is significantly inhibited. This forces more aggregate transport to the film, changing the behavior of the system. Again, the scaled system is not the same as the system under study in the laboratory.

The third and final item making a scaled experiment unacceptable is the assumption of pseudo-zeroth order conditions. If the depth of the 1 cm-diameter cell is reduced to 0.05 cm, with a concentration of 2.51 mg/ml, the total mass of monomer in solution is $1.25 \cdot 10^{-4}$ g. A film grown to a thickness of 0.15 μm, with the same diameter as the solution chamber and, having a density of 1.41 g/cm^3, consumes a mass of $2.1 \cdot 10^{-5}$ g of monomer, more than 10% of the original amount of monomer in solution. Clearly this would violate the assumption that the bulk and average concentrations do not change during the course of the experiment.

In summary, the system under study is extremely complicated. When the experiment is scaled down to suppress convection it no longer conforms to the necessary assumptions used to describe and model the system in the laboratory, nor does it any longer operate under pseudo-zeroth order conditions. This makes it impossible to use the scaling approach to perform convection-free experiments in the laboratory rather than in space. It has been shown, however, that reductions of four or five orders of magnitude in the g-vector are sufficient to achieve a convection free environment in this system. These conditions are readily achievable on the space shuttle and on space station, making them ideal locations for this research.

Microgravity Thin Film Growth Rate Equation

The film growth rate in microgravity is governed by two factors: the reaction kinetics for thin film growth and, if mixing is inadequate due to suppressed convection, the diffusive replenishment rate of nutrient monomer to the growing film. The starting point for deriving such a diffusive/kinetic rate equation is linking diffusive mass flux to the kinetically driven film growth process. The kinetic rate equation for thin film growth has been solved (*4*, *7*):

$$\frac{dl}{dt} = k \cdot I_o \cdot e^{-\varepsilon \cdot l} \cdot C_i^{0.5}. \tag{2}$$

Note that l is the thickness of the film at time t, k is the kinetic rate constant, I_o is the UV light intensity incident on the sample chamber, ε is the extinction coefficient of the film and C_i is the concentration of monomer in solution **at the interface**. It is important to note that the concentration of interest is the interfacial concentration, not the bulk solution concentration, which can be different in diffusion-controlled mass transport conditions.

Simultaneously with this kinetic activity removing mass from the interface is the already-mentioned diffusive transport providing mass to the interface. The change in concentration due to diffusive feeding of a depleting interface is

$$\Delta C = \Phi \left(\frac{t}{\pi \cdot D} \right)^{\frac{1}{2}} \cdot e^{-(x-x')^2/4Dt} - \frac{\Phi |x-x'|}{2 \cdot D} \text{erfc} \frac{|x-x'|}{2\sqrt{D \cdot t}} \quad (8).$$ Defining the terms for this equation, Φ is the mass flux from solution into the interface at coordinate x', x is the coordinate at which the concentration depletion is being evaluated, D is the diffusion coefficient of DAMNA in DCE, and t is the time over which the mass flux has occurred. For convenience, the location of the concentration sink, the film growth interface, is set to zero. Additionally, x can be set to zero since the conentration depletion region of interest is also the growth interface. Hence the negative term goes to zero, while the exponential term goes to one. This leaves $\Delta C = \Phi \left(\frac{t}{\pi \cdot D} \right)^{\frac{1}{2}}$.

Knowing that the mass flux is related to the film growth rate through the film density, ρ, by $\frac{dl}{dt} = \frac{\Phi}{\rho}$, and rewriting the change in concentration as $\Delta C = C_o - C_i$, where C_o is the bulk concentration and C_i is the interfacial concentration, leads to a diffusively governed equation for the rate of change of thickness, or growth rate, of the film:

$$\frac{dl}{dt} = \left(\frac{\pi \cdot D}{t} \right)^{0.5} \cdot \frac{C_o - C_i}{\rho}. \quad (3)$$

Setting Equations 2 and 3 equal to each other allows the interfacial concentration to be solved as a function of t and l for any set of initial conditions I_o and C_o, $\left(\frac{\pi \cdot D}{t} \right)^{0.5} \cdot \frac{C_o - C_i}{\rho} = k \cdot I_o \cdot e^{-\varepsilon \cdot l} \cdot C_i^{0.5}$. By rearranging this equation and using the quadratic formula, a solution is obtained for the interfacial concentration as a function of the two initial conditions, C_o and I_o, film thickness and time:

$$C_i^{0.5} = \frac{(-k \cdot I_o \cdot e^{-\varepsilon \cdot l}) \pm \left((k \cdot I_o \cdot e^{-\varepsilon \cdot l})^2 + \frac{4 \cdot C_o}{\rho^2} \cdot \left(\frac{\pi \cdot D}{t} \right) \right)^{0.5}}{\frac{2}{\rho} \cdot \left(\frac{\pi \cdot D}{t} \right)^{0.5}}. \quad (4)$$

Before proceeding, it is convenient to simplify Equation 4. Ground control experiments for the space experiment produced the result shown in Table I. These can be used to evaluate the square root term in the numerator of Equation 4. In all cases from $t=0$ to $t=3600$ seconds, the kinetic term on the left is at least three orders of magnitude less than the diffusive term on the right under the intensity and concentration growth conditions used in this experiment. The kinetic term can then be ignored for the purpose of this analysis and Equation 4 can be rewritten as $C_i^{0.5} = C_o^{0.5} - (k \cdot I_o \cdot e^{-\varepsilon \cdot l}) \cdot \left(\frac{\rho^2 t}{4 \cdot \pi \cdot D} \right)^{0.5}$. This relationship for the interface concentration can be substituted into Equation 2 for the microgravity kinetic rate equation, yielding a diffusive/kinetic rate equation for thin film growth applicable to microgravity:

$$\frac{dl}{dt} = k \cdot I_o \cdot C_o^{0.5} \cdot e^{-\varepsilon \cdot l} - (k \cdot I_o \cdot e^{-\varepsilon \cdot l})^2 \cdot \left(\frac{\rho^2 t}{4 \cdot \pi \cdot D} \right)^{0.5}. \quad (5)$$

This differential equation was integrated using the Runge-Kutta technique with initial conditions $l=0$ at $t=0$. This returned a set of data points for t, l that were then curve-fitted to produce a microgravity growth rate equation, $l = -2.60 \cdot 10^{-13} \cdot t^2 + 5.01 \cdot 10^{-9} \cdot t + 6.56 \cdot 10^{-8}$, shown graphically in Figure 4.

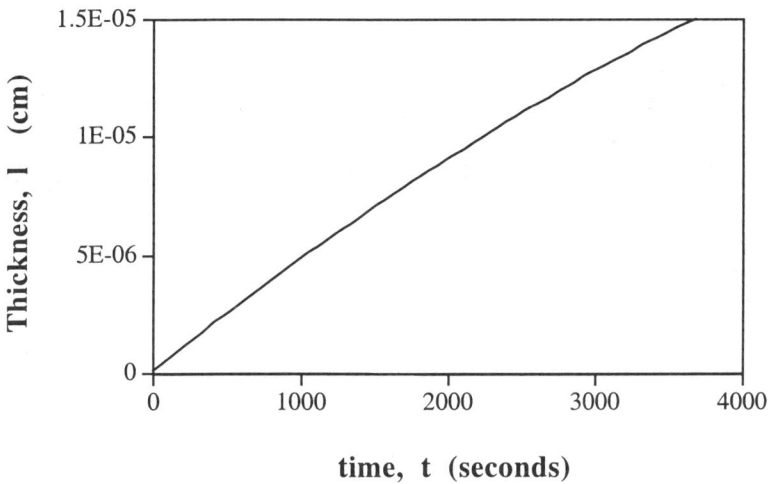

Figure 4. Film thickness as a function of time in microgravity. (Reproduced with permission from reference 4. Copyright 2000.)

Table I. Experimental conditions for ground-control experiments

Parameter	Value
k	$6.54 \cdot 10^{-10}$ (cm^{-1}·mW/cm^2·(mg/ml)$^{0.5}$·sec)$^{-1}$
I_o	5.1 mW/cm^2
ε	30,440 cm^{-1}
l	$(1.50\pm0.09)\cdot 10^{-5}$ cm (0.15 μm)
D	$(2.14\pm1.07)\cdot 10^{-5}$ cm^2/sec
ρ	$(1.41\pm0.04)\cdot 10^3$ mg/cm^3
C_o	2.51 mg/ml
t	3600 sec

The values for the extinction coefficient, ε, the diffusion coefficient, D, and the density, ρ, of the films were determined in our laboratory during the course of these experiments (4). The diffusion coefficient, in the range of 10^{-5} cm^2/sec, is in line with the values expected for molecules this size. The error term, however, is large. The reason this number was accepted and reported with this large error is the lack of need for more precision. Prior to yielding to the impulse to pursue these measurements to greater accuracy, an evaluation was made of the effect of the error on the derived model. Using only the mean answer, D=$2.14\cdot 10^{-5}$ cm^2/sec, the final thickness calculated by the model is 0.148 μm. Including the negative error term, or D=$1.07\cdot 10^{-5}$ cm^2/sec, the answer changes only slightly to 0.147 μm. Using the positive error term does not produce a change that can be resolved with three significant figures, yielding a thickness of 0.148 μm. With this information the decision was made to accept the diffusion coefficient with the error as is.

Microgravity Thin Film Growth Experiment

A microgravity experiment was carried out on STS-69 in September 1995. Samples were prepared using HPLC-grade 1,2-dichloroethane and DAMNA monomer purified by column chromatography (9). This experiment used the Polymer Thin Film Growth (PTFG) hardware, shown in Figure 5. It features a long-wave ultraviolet fluorescent bulb for polymerization. The sample cells, or chambers, are arrayed around and facing the fluorescent UV lamp, which runs down the center of the assembly. The DC/AC inverter converts the DC battery power into AC power to drive the fluorescent lamp and the conduit protects the wires leading to the far end of the bulb. The three chamber retaining bars each hold 5 sample chambers, providing a total of 15 samples for the experiment. The entire assembly is housed inside a polished aluminum cylinder, closed off at the end with aluminum caps. The UV lamp intensity was difficult to measure directly since it puts out a band of light rather than a single frequency, and the absorption of the light, and hence the reaction rate, of the monomer is wavelength dependent. Therefore the effective lamp intensity was determined by growing a set of films in the laboratory for one hour, then rearranging eq 2 and calculating the effective intensity based on the film thickness. This exercise produced a value of 5.1 mW/cm^2. Time and resource constraints of the project precluded measuring the UV lamp intensity on orbit. Fluorescent lights are used extensively on the space shuttle, however, and no performance degradations are experienced in them. Therefore we assume the UV lamps had the same intensity on orbit as they do in the laboratory. A UV intensity measurement capability is being incorporated into the flight hardware for subsequent experiments.

The results from the flight experiment are shown in Table II. The ground-control experiment results are shown in Table III. The sample from flight cell#4 was missing when the films were analyzed post-flight. The whereabouts of that space-flown window and film are unknown to this date, but it is believed they became souvenirs.

Film thicknesses after growth were determined using UV/Vis spectroscopy at 364 nm. The films were removed from their growth chambers and placed in the path of the UV beam, which was slightly smaller in diameter than the film being measured. Thus an average film thickness was obtained for each sample, irrespective of film thickness non-uniformities.

Discussion

The most notable aspect of the microgravity work is the discrepancy between the thickness of the film actually grown in space and the thickness of the films predicted by the diffusive/kinetic microgravity growth rate model. The ground-grown control films grew to 0.150 μm and the model predicted that the

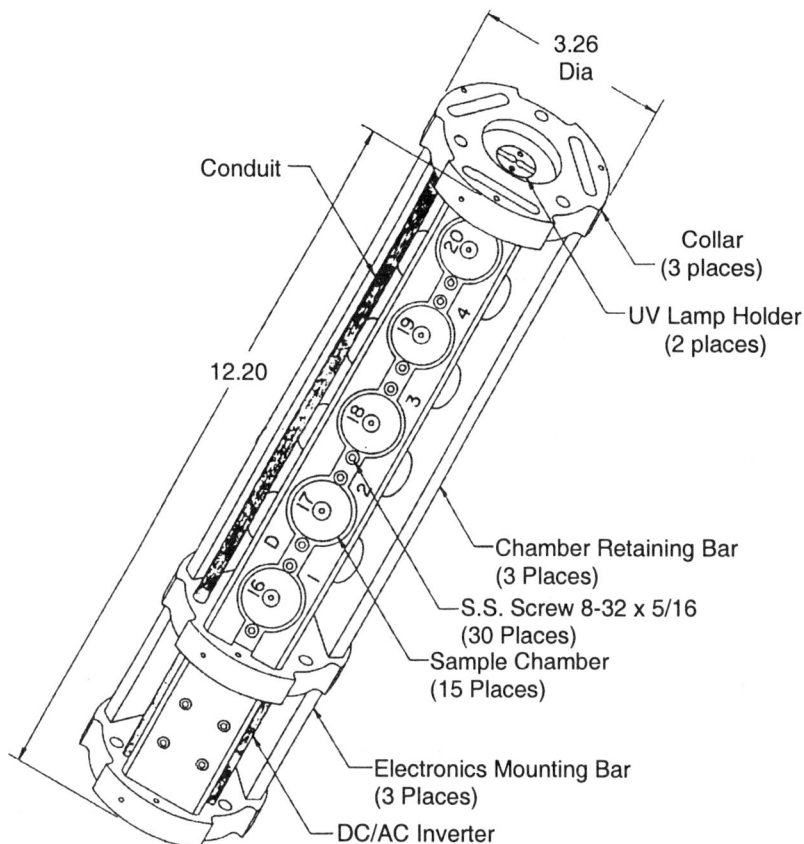

Figure 5. Microgravity polymer thin film growth (PTFG) hardware. (Reproduced with permission from reference 4. Copyright 2000.)

Table II. Results from the microgravity flight experiments

PDAMNA Thin Film Flight Samples	
Cell #	Thickness (μm)
2	0.104
3	0.103
15	0.100
4	Missing
	0.102±0.004

Table III. Results from the microgravity flight ground-control experiments

PDAMNA Thin Film Ground-Control Samples	
Cell #	Thickness (μm)
68	0.154
41	0.157
61	0.146
31	0.141
	0.150±0.009

space-grown films would grow to 0.149 μm. But the actual thickness of the space films was only 0.102 μm. Two factors are seen as contributing to this situation: the presence of impurities in solution and the polymerization/aggregation taking place in solution.

Solvents are never 100% pure and these impurities are well-known culprits in reducing diffusive nutrient flux to the growth interface in convection-free solution growth systems. Even though this is not a crystal growth system, the role played by impurities in retarding diffusive nutrient transport to the interface is the same.

The second factor reducing the film growth rate is believed to arise from the polymerization taking place in solution, which leads to aggregation of the nearly insoluble polymer chains. The aggregates are known to reach sizes on the order of 100 nm in less than one hour (4). This is a volume of approximately $5 \cdot 10^{-16}$ cm^3. With a measured density of 1.41 g/cm^3 and a molecular weight of 244 g/mol, a quick calculation reveals that a single 100 nm-diameter aggregate particle ties up almost 10^7 monomers. This can represent a meaningful local reduction in particle density. Furthermore, it has already been pointed out that the UV penetration depth is only about 0.3 mm into the solution, meaning that all of this number particle reduction occurs immediately adjacent to the interface. With all of this depletion depriving the interface of nutrient, it is easy to see that the growth rate can be seriously retarded in a convection-free environment. Further research will quantify bulk solution polymerization kinetics, providing quantitative insight into the mass tied up by aggregation, perhaps enabling the development of a more accurate model of the overall system.

Conclusions

This experiment appears to be a dramatic illustration of the role played by polymer aggregation in solution in a quiescent environment. However, while the data are very precise, with good uniformity in thickness from film to film, clearly more data points need to be obtained before meaningful and convincing conclusions can be made. To this end more space experiments are being planned. Work is in progress for one more space shuttle experiment, hopefully in 2001, and an additional set of experiments on the International Space Station, currently planned for 2004. With these data it may be possible to make decisive conclusions regarding the growth rate of films in microgravity. Also, eventually a rate law for polymer formation in solution will be obtained. It is hoped that this rate law, combined with knowledge of the aggregation behavior of that same polymer in solution, can be combined with the film growth rate law and the diffusive transport equations to generate a complete and accurate rate law for film growth in microgravity.

References

1. Paley, M. S.; Frazier, D. O.; Abdeldeyem, H. C*hem. Mat.* **1994**, *vol 6*, 2213-2215.
2. Paley, M. S.; Frazier, D. O.; Abdeldeyem, H.; Armstrong, S.; McManus, S. P. *J. Am. Chem. Soc.* **1995** *vol 117*, 4775-4780.
3. Frazier, D. O.; McManus, S. P.; Paley, M. S.; Donovan, D. N. U.S. Patent 5,451,433, 1995.
4. Carswell, W. E. Ph.D. dissertation, The University of Alabama, Huntsville, AL, 2000.
5. Frazier, D. O.; Hung, R. J.; Paley, M. S.; Long, Y. T. *J. Cryst. Gr.* **1997** *vol 173*, 172-181.
6. Brown, R. A. in *Materials Sciences in Space: A Contribution of the Scientific Basis of Space Processing;* Feuerbacher, B.; Hamacher, H.; Naumann, R. J., Ed.; Springer-Verlag: Berlin, 1986, pp 55-92.
7. Paley, M. S.; Armstrong, S.; Witherow, W. K.; Frazier, D. O. *Chem. Mat.* **1996** *vol 8*, 912-915.
8. Carslaw, H. S.; Jaeger, J. C. *Conduction of Heat in Solids*; Clarendon Press: Oxford. 1975; 263.
9. Paley, M. S.; Frazier, D. O.; McManus, S. P.; Zutaut, S. E.; Sanghadasa, M. *Chem. Mat.* **1993**, *vol 5*, 1641-1644.

Chapter 5

Polymerization in Microgravity On-Board STS–57, STS–63, and STS–77

Kenneth G. Brown[1], Karen S. Burns[1], Jo Ingram[1], George M. Wood[2], and Billy T. Upchurch[3]

[1]Department of Chemistry and Biochemistry, Old Dominion University, Norfolk, VA 23529
[2]STC Catalysts, 10 Basil Sawyer Drive, Hampton, VA 23666
[3]Langley Research Center, National Aeronautics and Space Administration, MS 493, Hampton, VA 23681

NASA-Langley Research Center has designed and built a polymerization module (PM) which allows for the routine synthesis of polymers onboard the space shuttle in Low-Earth-Orbit (LEO). This module fits inside a Commercial Refrigerator Incubator Module (CRIM) which is designed to fit in a locker in Spacehab®, a commercial laboratory facility located in the shuttle cargo bay. The PM was flown on three shuttle flights: STS-57 in June of 1993, STS-63 in February of 1994 and STS-77 in May of 1996. The monomers that were flown include miscible and immiscible systems such as: polyethyleneglycol 200 monomethacrylate, 3-Methacryloxypropyl-pentamethyldisiloxane, 4-vinyl pyridine, 1H, 1H Perfluorooctyl Methacrylate and methyl methacrylate This paper will discuss the design of the experimental package, its' on-orbit performance, the appearance of the polymer samples upon retrieval, the permeability, density and concentration profiles where possible.

INTRODUCTION

The near-zero gravity environment attainable aboard the U.S. space shuttle presents an attractive opportunity for the preparation of materials with properties that are impossible to achieve with ground-based systems. Materials which increase in molecular weight and/or size upon formation, such as polymers and crystals, are affected by gravitational forces since their density changes upon formation. Removal of the gravitational field would remove the density dependence. Molecular systems that naturally separate in a normal gravity environment also offer the opportunity for the formation of new materials when the gravitational stress is relieved on-orbit. Previous low-gravity experiments (*1,2,3*) involving the crystallization of large proteins, seem to indicate that it is possible to obtain larger crystals with a greater amount of order than is possible in a normal gravity situation. Hard-sphere colloids have been found to crystallize almost exclusively in hexagonal close-packed planes under microgravity conditions rather than the mixture of face-centered cubic and hexagonal close-packed planes observed in normal gravity conditions (*4*).

Polyacrylamide gels formed in orbit (*5*) exhibit properties that are different from those formed in a ground-based laboratory. Other polymer experiments done in Drop Towers, parabolic flights and on-board orbital platforms (*6,7,8,9,10*) show that the properties of the polymeric materials are quite different from their ground-based prototypes. Further evidence for the gravitational dependence of polymerization processes has been obtained from studies on gel polymerization under high centrifugal forces (*11,12*). It is evident that the gravity dependent convection mechanisms involving heat and mass transfer are important in the synthesis of polymers.

The potential for affecting the order of molecular systems, by reducing or eliminating convection processes in LEO, holds promise for the synthesis of large molecular systems such as polymers with novel properties. Synthesis under near-zero-G conditions could result in an increase in the amount of microcrystalline arrays in polymers. The presence of such order would significantly affect the properties of such systems. The permeability of these materials to gases would be affected, if order is increased, with an overall net decrease in the permeability but with the possibility of enhanced selectivity for one gas over another. The ability to control this process could lead to the design of materials for separating specific gases. This technology should be of great advantage to a variety of industries that depend upon polymers either as barrier materials or as materials that could selectively transport particular gases such as oxygen or hydrogen.

This paper describes an experimental package that was flown, as a joint NASA/commercial venture, onboard an independent laboratory that was designed for placement in the cargo bay of the shuttle. McDonnell-Douglas® and Spacehab® Inc. developed this laboratory, known as Spacehab® for the Commercial Development of space. It is designed to hold several experiments at one time in a controlled environment in which the astronauts can safely work. We will discuss the design and operational behavior of the packages that were flown aboard STS-57, STS-63 and STS-77 in June 1993, February 1994 and May 1996 respectively. Testing of the package including the engineering requirements that had to be met has been described elsewhere (*13*). The objectives of the experiment were to observe differences in density distribution, homogeneity, and gas permeation in the on-orbit prepared samples compared to samples prepared identically and simultaneously in a controlled ground experiment.

The physical properties and the composition of some of the polymeric materials will be discussed. Permeability to oxygen and hydrogen and density were determined for all of the materials that did not appear to have significant voids caused by bubble formation. All of the samples were in the shape of cylindrical rods approximately 1.5 cm in diameter and 6 cm long. The composition of all of the rods was determined along the rod axis. Both physical properties and composition were compared to samples that were prepared in parallel under normal gravity conditions. The samples that are to be described are limited to the non-commercial materials. Those that were prepared by the commercial partner Paragon® Optical have been previously described (*14*).

Experimental

Materials and Analysis

All of the materials were used as obtained from the manufacturer except for the removal of inhibitor from the monomers that had been inhibited. All of the materials used for each flight, except for proprietary experiments conducted by Paragon® Optical, have been described elsewhere (*15,16,17*). Borosilicate glass test tubes, 18mm x 150mm, which had been silated with a 5% solution of trimethylchlorosilane in toluene, were used for every sample in all flights. Magnetic stirrers, which were placed inside the test tubes that contained solutions that required stirring, were Teflon coated. All of the materials used were carefully deoxygenated prior to use and kept under a nitrogen environment throughout the process of sample preparation. Every monomer was obtained from samples with the same lot numbers. Enough sample was prepared to ensure that both flight and ground samples came from the same batch. All preparations were carried out at 4 °C to ensure that no polymerization could occur prior to orbital insertion. After the monomers were placed in the test tube a chorolastic® septum was inserted into the test tube and forced downward until the bottom of the septum came into contact with the liquid to ensure that there would be no void volume present. During this process the volume between the solution and the septum was under constant aspiration. The tubes were then capped with a polyethylene cap and held at 4 °C prior to placement into the PM. A typical tube, without magnetic stirrers, is shown in figure 1.

Figure 1. An assembled test tube, post-launch, showing the polyethylene cap, the chorolastic septum and the polymerized sample. This tube contained the immiscible monomers, polyethyleneglycol 200 monomethacrylate (PEG-200) and 3-Methacryloxypropyl-pentamethyldisiloxane (disiloxane).

The permeability of the polymers was determined using a 180° magnetic sector mass spectrometer. The calibration of the instrument and the completee procedure has been described elsewhere (15,18). Thin slices were cut from the prepared rods using a Buehker® Diamond-coated saw. Densities were determined using standard buoyancy technique on thin slices. Composition of the rods while still in the test tubes was determined using Raman Spectroscopy. Cross-sectional composittions of slices of rods were also determined with Raman Spectroscopy as wel as Infrared photoacoustic spectroscopy. Thermal properties of the samples was determined using a Perkin-Elmer® Differential Scanning Calorimeter (DSC).

Instrumentation

The hardware for the experiment consisted of the PM and a CRIM, (essentially a thermoelectric heater/cooler) manufactured by Space Industries® Inc.. Only one wall of the CRIM is active; good thermal contact with that wall was insured using a turnbucke device. The PM was inserted into the CRIM, which provided the environmental control for the PM and could be operated between 4°C and 40°C. The entire package, PM plus CRIM, was then installed into a locker in the Spacehab® module locatd in the cargo bay of the shuttle. The installation into the locker occurred, for each flight, 24 hours before launch and involved less than a fifteen-minute power loss to the package. All power to the package was provided by the buss on-board Spacehab® which was linked to the main shuttle systems. All instructions for temperature ramping, control and stirring motor activation could be entered via a keypad on the front of the CRIM. All of the instructions were stored in memory provided in the CRIM.

The PM was constructed from an aluminum block into which wells were drilled to a depth that could accommodate 18mm x 150mm borosilicate glass test tubes. Each PM contains 28 test tubes in a 7 x 4 array. An o-ring was placed in the bottom of each hole to help isolate the tubes from lift-off vibration. Each hole was drilled to be slightly larger than the tube so that an additional o-ring was placed around each tube to ensure that each tube would be firmly held in place and further isolated from vibration. Once all of the test tubes were placed in the PM, an aluminum cover was bolted into place, sealing the test tube compartment. This cover was fitted with a chorolastic® seal that provided both vibration isolation and a level of containment for the materials in the test tubes. When this was completed, another cover, designed to help the PM make thermal contact with the walls of the CRIM, was put in place. This last cover was also fitted with a chorolastic® seal providing another level of containment. In all, there were three levels of containment, the two covers and the chorolastic® septum, which had been inserted into each test tube. Additionally, the PM was fitted with several connectors for the attachment of thermocouples. Upwards of 9 thermocouples, extending from the CRIM, were affixed to the PM. Each thermocouple was read at five-minute intervals providing an in-depth temperature profile of the PM both prior to launch and during the orbital phase. The temperature data was stored in memory provided by the CRIM along with the internal CRIM temperature readings. When the entire package was recovered the thermal data could be downloaded and stored in an Excel® spreadsheet. The temperature profile for the mission could be determined for the entire package. Additionally, wires attached to the stirrer motors were attached to the CRIM, which provided for their activation and deactivation.

The stirrer motors were located immediately adjacent to the PM on two sides of the PM. The spinning action was provided by a magnet located on the rotating arm of each motor. It was determined that only three test tubes, located in the outside row of the PM would be affected by the magnetic field providing a field at the center of the tube that would be sufficient to effectively spin the magnetic stirrer. As a result only six test tubes, three on each side, contained a magnet to provide for stirring the solution. In the row immediately behind these tubes were tubes containing the same samples but without magnetic stirrers. On both STS-63 and STS-77 only one of the in-flight CRIMs was equipped with stirrer motors. An artist's cut-out view of the assembled package showing the

PM with the tubes in place is shown in Figure 2. When the entire package was mounted in the shutttle, the long axis of the test tubes carrying the samples was perpendicular to gravity normal (i.e. they were laying on their sides). The turnbuckle mechanism which provides for good thermal contact with the active wall is at the right of the figure. The PM with the stirrer motors in place is not shown.

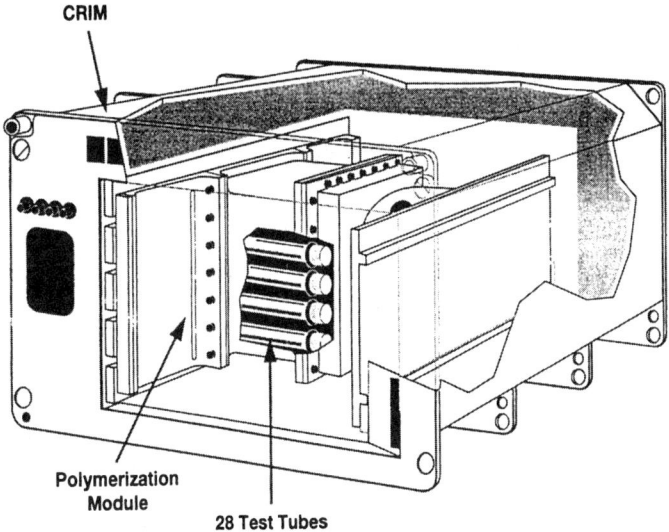

Figure 2. A cut-out view of the CRIM showing the PM loaded with test tubes.

Procedure

A typical mission profile, from sample preparation to after-launch recovery will now be described. When the test tubes were loaded, inserted into the PM, and all thermocouple and stirring motor connections were completed, the PM was inserted into the CRIM approximately 24 hours prior to scheduled pick-up by Spacehab® personnel. At this time, the expected mission profile was loaded into CRIM memory and the CRIM was set to hold the samples at 4°C prior to pickup. The ground-based samples were treated in the same manner utilizing refrigerator-incubators in the laboratory. The package was removed from the laboratory to the launch pad utilizing a battery cart. Overall power loss when switching from laboratory power to the battery cart was less than 5 seconds. At the launch pad the package had to be disconnected from the battery cart and lifted into Spacehab®. Typically the interruption was less than five minutes. Launch was, in general, twenty-four hours later. Entrance into Spacehab®, by the mission specialist, occurred four hours after orbital insertion had been achieved and the program on the CRIM was activated. The ground packages were, then, started simultaneously with the on-orbit packages. All three missions were about 10 days duration. The experiment was halted by the mission specialist 24 hours prior to scheduled de-orbit. and ground packages were halted at the same time. At this point the temperature in the CRIM and in the ground packages was allowed to return to ambient temperature without control. The CRIMs were retrieved from the Spacehab® facility approximately 4 hours after landing. All of the packages were returned to NASA-Langley in Hampton, VA for opening, photographing and retrieval.

The preprogrammed temperature profile loaded aboard the CRIM was designed to hold the samples at 4°C until the remainder of the program was activated on-orbit by the mission specialists. At this time a temperature ramp to 40 °C (0.1 °C/min) was begun in both the on-orbit modules and the ground facility. During STS-63 and STS-77, stirrer motors were also activated. The temperature rise was determined, from ground-based experiments, to take six hours. At the end of the six hour period the CRIM and the ground-based experiment began to control the temperature at 40°C and the stirrer motors were turned off. The polymerization was, then, allowed to continue at 40 °C until 24 hours prior to scheduled de-orbit. A typical temperature profile for one of the PMs is shown in Figure 3. In all of the missions the temperature contol occurred without an error.

Results and Discussion

The monomers flown on all three missions consisted of those that were completely miscible with one another as well as those which demonstrated extensive immiscibility. The miscible systems were assembled for their previously demonstrated permeability to atmospheric gases. This was in keeping with the original intent of the project and the objective of our commercial partner, Paragon® Optical. In addition some of the monomeric mixtures were modified to contain metal atoms which would be incorporated into the polymer. The latter was an effort to enhance the permeability of the materials to specific gases and to determine the effect of low-gravity on the distribution of the atoms within the polymeric material. In STS-57, we observed that the immiscible monomers, 4-vinyl pyridine and 1H, 1H Perfluorooctyl Methacrylate experienced unexpected mixing while undergoing low-gravity polymerization. In subsequent flights, we further explored the mixing including the addition of stirring to enhance the mixing. The materials formed from immiscible monomers contained numerous cavities and were not suitable for either density or permeability measurement, but their composition profiles could be determined.

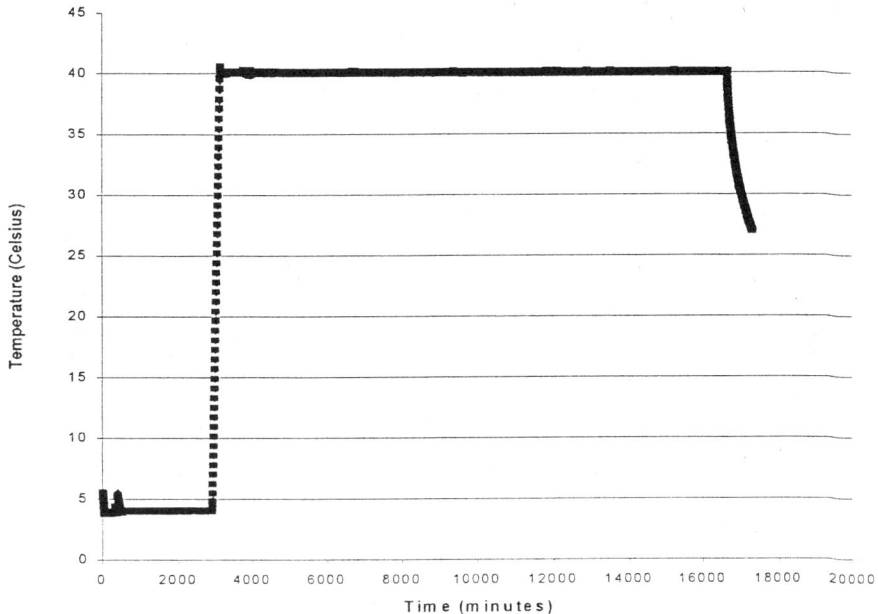

Figure 3. Temperature profile of a typical mission. The steep rise in the temperature corresponds to a six-hour period during which the temperature is ramped from 4 °C to 40 °C. The slow decline in temperature at the end of the 40 °C period represents the packages unaided cooldown.

Density and Permeability

Upon visual inspection the homopolymers, formed from the miscible monomers, appear to be almost identical for both flight and ground samples. The density measurements revealed that any differences in density between flight and ground were very small. In figure 4 we show the density profile for a rod formed from methyl methacrylate which contained 0.25% Palladium acetonyl acetonate with 2.0% ethylene glycol dimethacrylate (EGD) as a cross-linker. This was sectioned into 25 very thin slices along the long-axis of the rod for both the flight and ground samples. The flight-synthesized (STS-63) rod is slightly more dense at the end of the rod furtherest from the septum. The ground-synthesized rod has a density profile which is more uniform. The same materials, without Pd, exhibited no density differences along the rod for either the flight or ground samples. These results are fairly typical for all of the systems where the density could be measured. Any differences between flight and ground were exceedingly small, density profiles were as in figure 4 or the reverse.

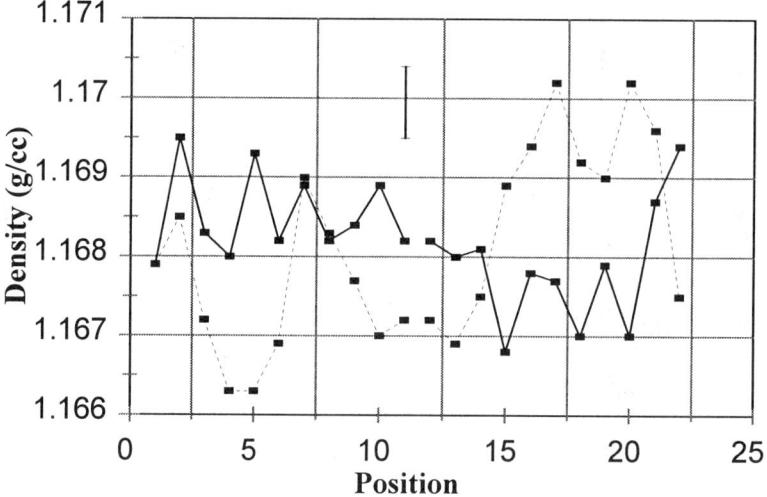

Figure 4. Density of Palladium containing rods from the second mission. The dashed line is the flight-synthesized sample, the solid line is the ground-synthesized sample. The vertical line represents the standard error. Position 0 represents the septum side of the rod, position 25 the bottom of the tube. The densities were determined at 25 °C.

The permeability, Dk, (where D is the diffusivity and k is Henry's Law constant) of the above samples to both hydrogen and oxygen were determined. The results for hydrogen are shown in table I. The Pd-containing flight-synthesized samples are more permeable to hydrogen than their ground-synthesized counterparts. This difference remains whether or not the measured permeability was on slices that had been obtained from the PM (prereduced) or after being exposed to hydrogen for 24 hours (postreduced). For the flight rods, the permeability in the prereduced state is greater at the end of the rod furtherest from the septum by about 18%, whereas, in the ground rod the trend is reversed and the difference is about 13%. The difference disappeared for the flight specimens and completely reversed for the ground synthesized samples in the postreduced state. The flight samples without palladium differ by only 10% end-to-end with the trend in the opposite direction of those containing palladium. The reduction reduces the permeability but maintains the trend. The ground-synthesized samples without Pd have the highest permeabilities with the end furtherest from the septum always exhibiting the highest permeability either before or after reduction.

Table I. Hydrogen permeability for Palladium containing materials. Mission STS-63

MATERIAL		FLIGHT (Dk)	GROUND (Dk)
Methyl Methacrylate	Top – pre	7.6	6.8
Ethylene glycol dimethacrylate	Top – post	8.7	6.6
Palladium acetonyl acetonate	Bottom – pre	9.3	5.9
2,2'-azobis-(2,4-dimethylvaleronitrile)	Bottom – post	8.7	7.6
Methyl Methacrylate	Top – pre	9.0	9.0
Ethylene glycol dimethacrylate	Top – post	8.9	8.6
2,2'-azobis-(2,4-dimethylvaleronitrile)	Bottom – pre	8.0	10.7
	Bottom – post	7.8	12.0

Note - Values of Dk should be multiplied by 10^{-11} ml/cm · s · mmHg. All reported values are the average of at least three trials with a stadard deviation of ± 0.05.
Top refers to the end of the rod nearest the septum.
post = postreduction, pre = prereduction
All hydrogen post reduction values were obtained after membrane was exposed to hydrogen gas flow at 4.5 sccm for 24 hours.

Trends in oxygen permeability were not as evident as for hydrogen. Part of the reason for this is the greatly reduced permeability for oxygen, 10% or less, of the value for hydrogen.

Composition

The flight systems consisting of miscible monomers exhibited no unique concentration profiles. The ground-based synthesis displayed no significant differences from the flight-based. However, the systems involving the immiscible monomers exhibited significant difference between flight and ground. In general, in the ground-based synthesis, the two layers remained separate with the position of the layers in the tube being defined by the orientation of the tube in the earth's gravitational field. Those synthesized in microgravity tended, when there was no mixing of the monomer,

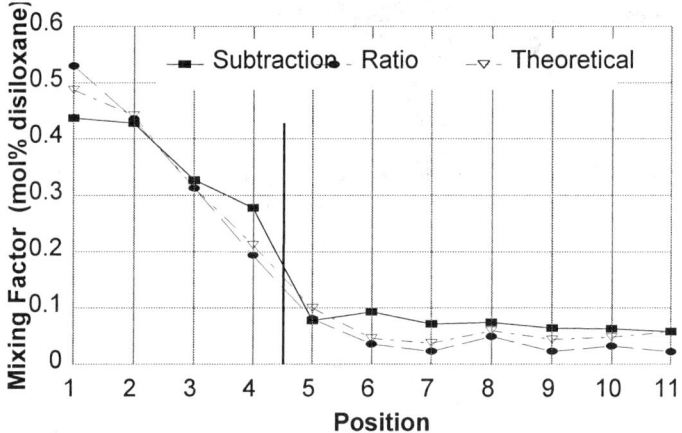

Figure 5. Concentration profile for the flight sample shown in figure 1. Position 1 is the end of the tube nearest the septum. The three lines represent the three different techniques that were used to determine composition. The concentration factor stands for mole% 3-Methacryloxypropyl-pentamethyldisiloxane.

to form multi-layered systems as each phase tried to assume a minimum free energy surface. In some cases, as shown in figure 5 there appeared to be a considerable amount of mixing between the phases, which was not in evidence in the ground-based system (figure 6).

The first observation of mixing of immiscible monomers was demonstrated in STS-57 in a system involving 4-vinyl pyridine and styrene. It was this system that led us to incorporate mechanical stirring in the next two flights for the immiscible monomers. The effect of stirring is not altogether clear at this moment. In the ground-based synthesis we did not observe any enhancement in the mixing process. In the flight sample, for the mixture described in the

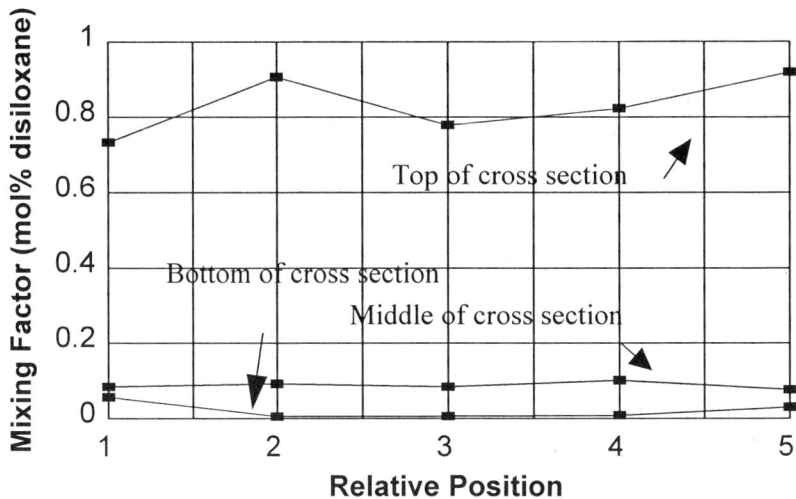

Figure 6. Concentration profile of the ground-synthesized materials which are the same as shown in figure 5. Position 1 is the end near the septum. The polymerization occurred with the tube on its side. There were two distinct phases. The three lines represent a sampling of a slice from one side of the tube to the other.

preceding figures, the stirred sample appears cloudy throughout the length of the test tube except for a small clear portion at the curved end of the tube which was unmixed PEG-200. The cloudy layer was, for the most part well-mixed in PEG-200 and 3-methacryloxypropyl-pentamethyldisiloxane (disilox). However, immediately around the stir-bar was a region that was approximately 80% disilox the remainder of the sample consisting of approximately 20% disilox

Another system flown consisted of polyethyleneglycol 400 methacrylate (PEG-400) and disilox. Minimal mixing was observed in this system flight or ground. The PEG-400 is a monomer that has a higher molecular weight than the PEG-200. In addition it has a higher viscosity, 59.5 ± 0.4 cP for PEG-400 versus 33.21 ± 0.07 cP for PEG-200. The surface tension of PEG-200 was 39.0 ± 0.6 dyne/cm and we could not determine the surface tension of PEG-400. For reference purposes the surface tension of the disilox was 22.0 ± 0.6 dyne/cm and the viscosity was 2.282 ± 0.008 cP. The only difference between the flight and ground samples, stirred and unstirred was the location of the almost pure phases. Some mixing was observed for the stirred flight sample.

Conclusions

The total number of samples that have been characterized by the teams at NASA-Langley and Old Dominion University is 60 for all three missions, with the remaining spaces being utilized by Paragon® Optical. We are on the upward slope of a learning curve. Some early conclusions are possible but they are by no means definitive.

The difference between flight and ground synthesized sample, while not extreme for polymers formed from miscible monomers, were significant. Whereas density differences were not readily apparent there were differences in permeability, although they were not always of the same extent between samples of the same chemical composition. In addition, some of the observed differences appeared to diminish considerably when the samples were exposed to normal treatment on earth such as annealing and grinding.

The polymers formed from the immiscible monomers seem to hold promise for the development of new materials that cannot be synthesized on the ground except, perhaps in the presence of a solvent. We are seeing a mixing that is driven, in large part, by surface tension differences. The benefit of mechanical stirring needs to be investigated further.

It is quite likely that differences between flight and ground synthesis could be better demonstrated if the processes could be carried out much more slowly. The maximum duration of the experiment was eight days. In actuality each mission was designed so that the polymerizations were carried out faster than that with the polymerization being essentially complete in three to four days. We felt that we needed to be ready for a decision to shorten the mission. An experiment of longer duration, such as on-board space station would be extremely desirable.

Acknowledgement

We would like to thank the NASA office of Commercialization of Space for funding and our colleagues at Paragon® Optical for their collaboration.

References

1. DeLucas, L. J.; et. al, *J. Crystal Growth*, **1991**, 110, 302-311.
2. Strong, R. K.; Stoddard, B. L.; Arrott, A.; Farber, G. K., *J. Crystal Growth*, **1992**, 119, 200 - 214.

3. Littke, W.; John, C., *J. Crystal Growth*, **1986**, 76, 663 - 672.
4. Zhu, J.; Li, M.; Rogers, R.; Meyer, W.; Ottewill, R.; Russel, W.B.; Chalkin P., *Nature* **1997**, 387, 883-885.
5. Vanderhoff, J. W.; ElAasser, M. S.; Micale, F. J.; Sudol, E. D.; Tseng, C. M.; Silwanowicz, A.; Sheu, H. R., In *Proc.Amer. Chem. Soc.*, Div. Polym Mater. Sci., **1986**, 54, 587 – 592.
6. Meier, M.; Pamperin, O.; Dittus, H.; Rath, H. J., in Abstr. Of VIII *Europ. Symp. On Material and Fluid Sciences in Microgravity*, ESA SP-333, **1992**, 384.
7. Muller, M.; Sturm, D.; Rath, H. J., in Proc. Of VIII Europ. Symp. On *Material and Fluid Sciences in Microgravity*, ESA SP-333, **1992**, 2, 895 899.
8. Abdurakhmanov, Sh. D.; Bogatyreva, L. G.; Briskman, V. A.; Levkovich, M. G.; Leontyev, V. B.; Lyubimova, T. P.; Mashinsky, A. L.; Nechitailo, G. S., In *Numerical and Experimental Modelling of Hydronamic Phenomena under Weightlessness*, Sverdlovsk, **1988**, 120 – 128.
9. Bogatyreva, L.; Briskman, V.; Levkovich, M.; Lyubimova, T; et. al., *Space Sci. Technol.*, **1989**, 43 – 47.
10. Abdurakhamonov, Sh.; Babskii, V.; Bogatyrev, l.; Briskman, V.; et. al., In *Gagarin Scientific Readings on the Astronautics and Aeronautics*, Nauka, Moscow, **1990**, 219 – 228.
11. Briskman, V. A.; Kostarev, K. G.; Lyubimova, T. P., *In Materials Processing in High Gravity*; Regel, L. L.; Wilcox, W. R., Eds.; Plenum Press, New York, **1994**, 185 – 192.
12. Briskman, V. A.; Kostarev, K.; Yudina, T., In *Centrifugal Materials Processing*; Regel, L. L.; Wilcox, W. R., Eds; Plenum Press, New York, **1997**, 247 – 255.
13. Applin, J.; Turner, C., *AIAA* 95-3786 September **1995**, 1-4.
14. Kulkarni, C. L., *AIAA* 95-3588 September **1995**.
15. Brown, K. G.; Burns, K. S.; Upchurch, B. T.; Wood, G. M., In *Polymeric Materials: Science and Engineering*, **1996**, 75, 294 – 295.
16. Brown, K. G.; Burns, K. S.; Upchurch, B. T.; Wood, G. M.; Applin, J. I. B.; Turner, C. P., In *Polymeric Materials: Science and Engineering*, **1996**, 75, 298 – 299.
17. Burns, K. S.; Brown, K. G.; Upchurch, B. T.; Wood, G. M., In *Polymeric Materials: Science and Engineering*, **1996**, 75, 296 – 297.
18. Upchurch, B. T.; Wood, G. M.; Brown, K. G.; Burns, K. S., In *NASA CP-3189*, **1992**, 1, 451 – 457.

Chapter 6

Production of Polyurethane Foams in Space: Gravitational and Vacuum Effects on Foam Formation

Samuel P. McManus[1,3], Francis C. Wessling[2,3],
John T. Matthews[1,3], Darayas N. Patek[1,3], Kazunori Emoto[1,3],
Douglas T. Howard[2,3], and T. Scott Price [2,3]

Departments of [1]Chemistry and [2]Mechanical and Aerospace Engineering, and [3]the Consortium for Materials Development in Space, The University of Alabama at Huntsville, Huntsville, AL 35899

Experiments have been conducted using Consort series research rockets to evaluate production of polyurethane foams both in a microgravity environment and in a space environment. The results of those experiments are discussed.

Introduction

Closed-cell Foams

Cellular polymeric materials with closed cells, which may be formed by incorporating gas bubbles into the matrix during polymeric cross-linking reactions, have widespread use for their insulating and structural properties (2,3). Because of their favorable strength-to-weight ratio, cellular materials enjoy numerous uses in aircraft and spacecraft. One can envision a variety of applications of foams in space. Because of their strength-to-weight properties and also because of volume-to-weight properties, a number of applications of polymeric foams that are manufactured in space can be envisioned. This chapter includes a discussion of our experience with foam formation in a microgravity environment and in space.

The Microgravity Environment

When considering formation of a foam in a microgravity environment, the shapes of the foam and its cells take on an interesting dimension because certain forces which affect them may be absent or reduced. For example, gravitational effects, which may compress and thus distort the cells and the gross structure, are substantially reduced. Also, surface tension, which is viewed as a main driver of cell shape, may be more evident in the absence of other factors.

For centuries, mathematicians have considered the shapes of fluid bubbles. This interesting story has been recently reviewed by Klarreich (4). Soap bubbles are the simplest and best studied form. Because of the availability of an inexpensive toy kit, many of us as children experimented with forming soap bubbles and observed that single soap bubbles formed as spheres. Theoreticians tell us this occurs because the bubble is adopting the minimum surface area. If one forms a series of liquid bubbles that join, it is nontrivial to predict the shapes of the cells. Based on minimum surface area considerations, Kelvin predicted a cell structure (Figure 1) that lasted for over a century as the preferred predicted core structure of complex liquid bubbles. In 1993 the Irish physicists Denis Weaire and Robert Phelan observed that certain chemical clathrates had structures with shapes that differed from predictions of the Kelvin model and decided to reconsider the problem theoretically. Their basis for reconsideration may have been flawed as complex chemical molecules typically adopt particular geometrical structures largely because of preferred bonding angles and not because they are trying to achieve the lowest minimum surface area. Nevertheless, upon the computation of the areas of many potential structures, a new minimum surface area structure was found (Figure 1) (4). To our knowledge there is no proof that the Weaire-Phelan structure is the final answer. Our results provide some insight on the structure for complex foams.

The initial experimental goal that we sought was to determine whether nearly perfect shapes would indeed form in a low gravity.

Space as a Reaction Environment

A goal that had a more practical end was production of polymeric foams in space for use in space. The space environment is considered a severe environment for the production and/or for the use of certain polymeric materials. First, consider the production of polyurethane foams, the cellular materials that our team chose to study. These foams are typically prepared by the reaction, catalyzed by base or by certain transition metal salts, of a di- or polyisocyanate with a prepolymer containing at least two reactive hydroxyl groups. The isocyanate and pre-polymer are mixed by stirring or vigorous shaking. Foam formation is carried out using a blowing agent, typically a volatile fluorocarbon, which becomes less soluble in the medium as the chain grows. The reaction exotherm leads to gasification of the fluorocarbon resulting in a 10-20 fold volume increase, and, in favorable cases, a homogeneous distribution of small cells in the resulting foam.

Imagine carrying out the above typical process in the space environment. First, if after initial mixing, the process is conducted without further stirring or shaking, in a low gravity environment, then gravitational effects are very different from those found in the laboratory (unit gravity). What are the effects of the reduced gravity levels on these reactions? From other studies (5,6), we anticipated that differences in surface phenomena and diffusion factors could affect mixing and thus reaction characteristics. We were also interested in the lack of gravity on the ultimate cell shapes, which might affect the foam's dimensional strength. However, from basic considerations, we were concerned about the effect of the low ambient pressure. With increasing altitude, one approaches a vacuum and thus foam processing in a near vacuum is a major concern. Intuitively one imagines the vacuum of space stripping the gaseous blowing agent from the creamy polymeric matrix before structural hardening occurs, which ultimately would protect the closed cell structure. Indeed, the engineering required to prevent this scenario proved to be the greatest technical challenge in producing usable polymeric foams in space.

Space as a Harsh Environment

There are elements of the space environment that make it harsh on materials that are employed in space. The presence of atomic oxygen and the ever-present ultraviolet rays of the sun are dominant. Some of the very best scientific data on the survivability of materials in space come from experiments conducted on NASA's Long-Duration Exposure Facility (LDEF). Summaries of those data (7,8) reveal that atomic oxygen may be a major problem for materials that deteriorate upon oxidation. Most dramatic among the LDEF samples was the erosion of exposed carbon surfaces owing to the loss of surface carbon as CO and CO_2. These results became design considerations.

Results and Discussion

Cell Structure of Polyurethane Foams

Photomicrographs of cross-sections of polyurethane foams do not lend unambiguous support for either the Kelvin or Weaire-Phalan structure. When slicing through a foam sample, many cells are included and many are dissected at angles that make analysis of an individual cell as a model for gross structure difficult. When good photomicrographs are analyzed, however, there appears to be reasonable support for the Weaire-Phalan structure as there are probably more pentagonal and hexagonal polygons included than others (Figure 2). However, many different polygons can be found as sides of some of the complex structures. This may indicate that there is an insufficient energy difference between the various polygonal structures for minimum surface energy to totally dominate with the rather viscous fluids that polyurethane foams represent as they form. But, clearly, the theorists are on track.

Gravitational Effects on Foam Formation and Foam Structure. High Gravity Effects

Since there is great interest in the structural integrity of cellular structures, one initial interest in the gravitational effects on foams was in the variation in cell shape as the gravity levels varied. From a cursory consideration of gravity effects, one would predict that the presence of gravity will tend to deform cells from their pristine geometrical shapes. However, this analysis is not straight forward since the pressure generated inside the cells by the blowing agent may overcome the forces of gravity thus leading to no observable gravitational effect. In fact, we knew that the gross structure of our foam was affected by gravity since we had to contain the foam mixture as it cured or else it flowed to a rounded thick sheet (pancake-like) on the lab benchtop. The absence of gravitational effects would have predicted formation of a foam ball spherical in shape.

While the polyurethane foam formulation we employed would flow to take the shape of its container, the cellular microstructure is not noticeably distorted by gravity (see Figure 3). Thus there must be a close balance, at the cell level, between the effects of gravity, which would work to deform or flatten the cells, and the pressure of the blowing agent that is exerted as an internal force on the cells that is equal in all directions. High gravity experiments, conducted in a centrifuge at ca. 100 gs, (1 g is equal to earth's gravity) did produce a noticeable

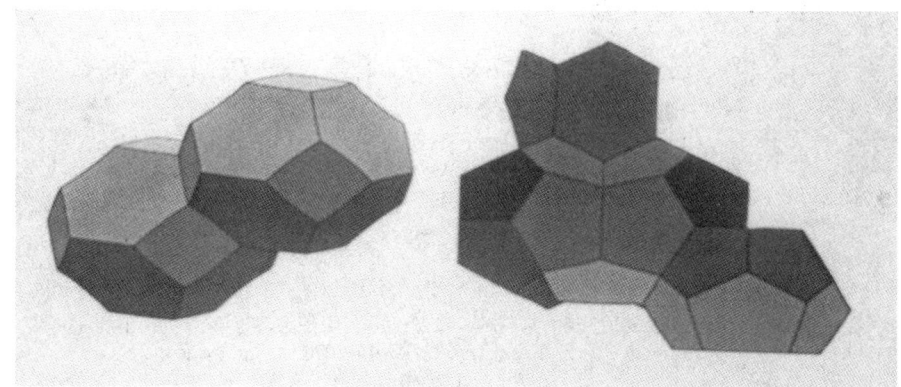

Figure 1. Models of the Kelvin (left) and Weaire-Phalan Structures (right) (4)

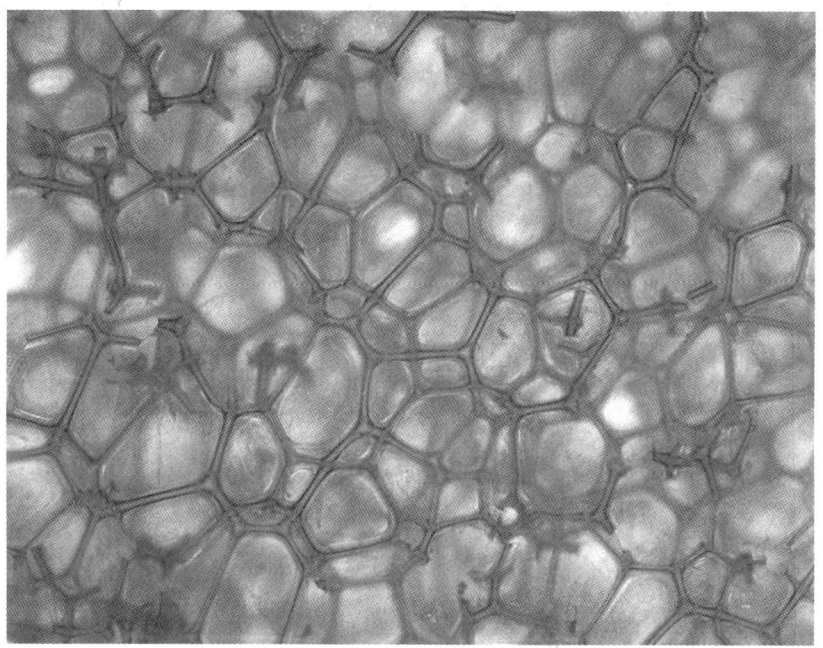

Figure 2. Photograph (X100) of a polyurethane foam cross-section.

gross difference in the shape of the resulting polyurethane foams (9). However, an evaluation of photomicrographs (Figure 3, top two pictures) of the foams revealed that the cells are not statistically different, owing to a variety of cell sizes. If one concentrates on the larger cells and the voids (not shown in the figure; voids are unusually large cells that sometimes form in poorly mixed foams), there is evidence of gravitational deformation. The high G levels also had an effect on mixing that affected foam formation at high gravity levels; some of the diisocyanate prepolymer (PAPI 94, Dow) tended to separate into pools (a centrifuge effect), which led to a low-quality foam.

The separation of diisocyanate due to high gravity effects on the mixture made a fair comparison of mechanical properties of the resulting foams unreasonable as the compositions of the foams were not identical. However one could readily see that the foams prepared in a vacuum, regardless of gravity level, were not closed-celled foams. The chlorofluorocarbon blowing agent ripped through cell walls as it was stripped from the forming foam.

Microgravity Effects

We were fortunate to have access to Consort–series research rockets, managed by our consortium, that were flown at White Sands Missile Range. These rockets allowed about seven minutes of experiment time with gravity levels near zero g. This experimental setup has been described in detail elsewhere (10). An experimental device was designed that would allow mixing and dispensing of the two-component polyurethane foam system just prior to the beginning of the microgravity phase of the rocket flight. A reaction mixture was chosen that (1) was compatible with the reaction apparatus and stable to the environmental conditions, including the rocket flight, (2) would provide a foam that would completely expand by the end of a seven-minute microgravity period, and (3) would provide a foam that was sufficiently cured within ten minutes after mixing to render it dimensionally stable enough to survive landing.

The mixing chamber included a motor-driven stirrer that mixed the components for 20 seconds prior to the expulsion of the mixture into an outer chamber where the foaming process was photographed. Unfortunately, experimental constraints of space experiments always include size, weight and power. With the constraints imposed on our experimental device with the research rocket program, we were forced to accept a stirring motor that provided good but not exceptional stirring. To form void-free foams, intimate mixing of the reaction components is required and is achievable with exceptional stirring. Without exceptional stirring, we were forced to abandon the goal of seeking the comparison of the physical properties, e.g. tensile strength, of a foam prepared in a microgravity environment with those of a foam prepared at unit gravity. We

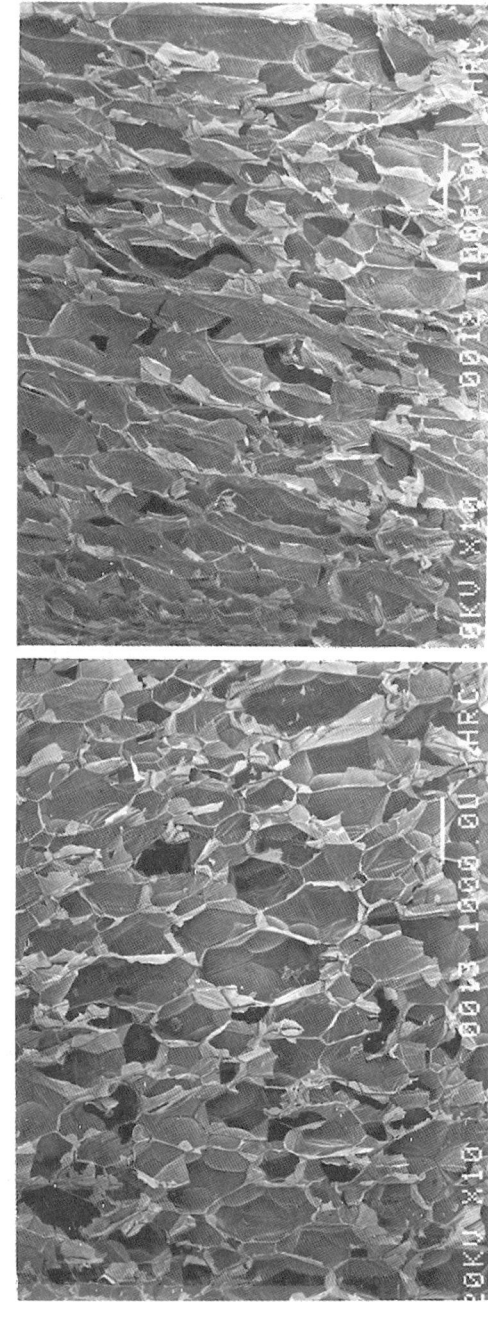

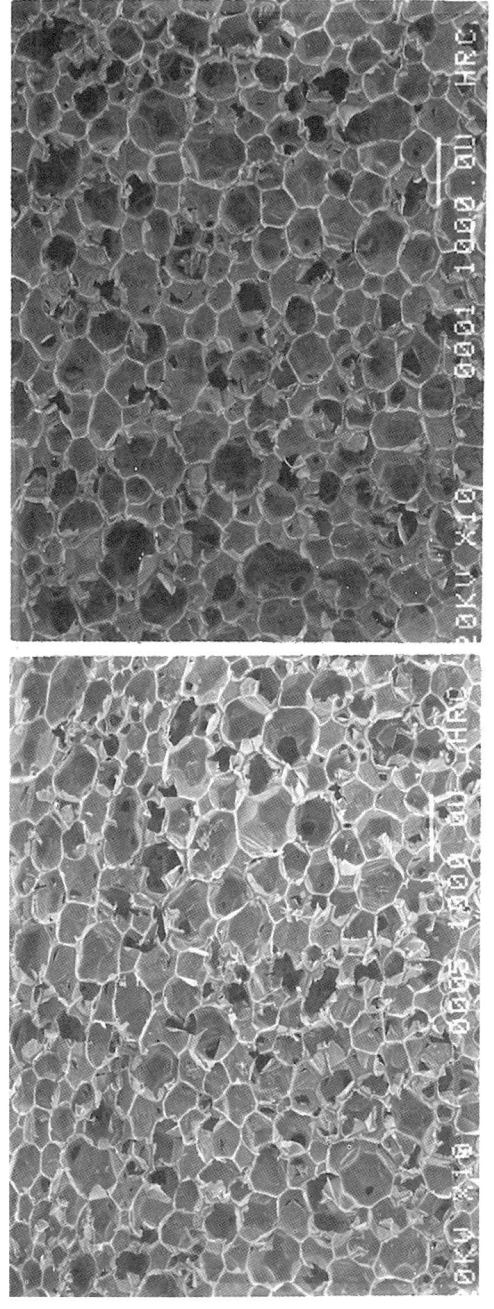

Figure 3. Photographs at a magnification X10 of polyurethane foams prepared under various laboratory conditions. Upper left: 1G, 1 Atm; upper right: high G, 1 Atm.; lower left: 1G, vacuum; lower right: high G, vacuum. Scale Bar = 1 mm.

still, however, desired to compare the gross effects of the microgravity experiment. To prevent having to deal with vacuum effects during this experiment, the experimental chamber was sealed from the rocket's external environment allowing the foam to form at approximately one atmosphere pressure.

As predicted, in the absence of significant gravity, the foam formed during the first Consort rocket flight, grew into a spherical ball. The process was photographed in stages allowing a record of stable foam growth. The resulting foam, which survived landing intact, has been analyzed (*11,12*). Figure 4 shows a cut-away of the ground-based sample prepared within the space hardware using all the same conditions except that the sample was allowed to flow into an aluminum cake pan to prevent it from pancaking onto the lab benchtop. Surface tension effects caused the foam formed during the Consort flight to hang as a spherical ball onto the exit cone. The spherical shape of the foam formed in microgravity is evident even in a cut-away of the sample, e.g. Figure 4 (right sample, e.g. left edge). Also, the effects on foam shape of the cake pan are evident in the ground-based sample (Figure 4, left sample).

The second most evident difference in the two samples is the gross comparisons of the large cells or voids. In each case, as pointed out above, the voids result from the incomplete mixing that was a limitation of the apparatus. The sample formed in space at 1 atmosphere contains voids that are nearly spherical, and there are many of them with similar sizes. The voids in the ground-based sample are irregular in shape and vary considerably in size. Another noticeable difference is the stress crack that resulted in the ground-based sample.

Figure 5 contains digital representations of cut-away portions of the two samples shown in Figure 4 (*13*). From a comparison of the cell microstructure (Figure 4, left comparisons), one can clearly observe that the microgravity sample has cells with a smaller standard deviation from the average cell size and that the average cell size is smaller. A more quantitative analysis of the digital representations reveals that the microgravity sample has about four times the number of cells. The digital depictions at the right of Figure 5 again reveal the gross differences in the cell voids that we discussed above.

Effects of Aluminum Particles. Dependence on Particle Type (*9*)

We had an interest in knowing whether aluminum coatings would improve the survivability of the foams in space. Rather than preparing the foams in a skin of metal or other material, we evaluated intimately mixing aluminum particles into the foam mix. We could find no previous mention of studies of the effects of aluminum particles on mixing or on ultimate foam properties. Interestingly, with

Figure 4. Photograph of cross-sections of polyurethane samples prepared in a microgravity environment at 1 atmosphere in a sounding rocket experiment (right) compared to its ground-based comparison (left).

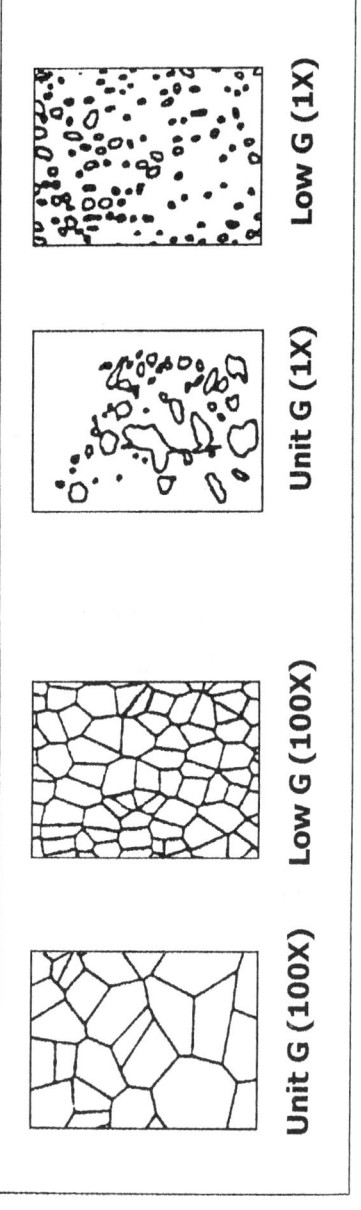

Figure 5. Digitized representations of polyurethane foams. The representations on the left represent comparisons of typical individual cell clusters at 100X magnification. Those on the right represent a comparison of the larger cells or voids that are readily observed without magnification.

two forms of small aluminum particles available, flaky (Reynolds Metals, 5xxx series alloys, particle size 3 microns to 100 microns on a side) and granular (Aluminum Co. of America, atomized powder, 2.5 microns to 20 microns on a side), different results emerged. The first notion that something different is happening is in the appearance of the product. A picture of comparison samples is shown in Figure 6. The granular aluminum made the normally cream-colored foam take on a gray color (Figure 6, left sample). The full foam sample prepared with the granular aluminum did not have the metallic appearance of the foam containing the same weight of flaky aluminum (Figure 6, right sample).

The only process difference noted when preparing the foams with aluminum was a 10 to 15° increase in the exotherm temperature for the foam. This increase in exotherm was most notable for the flaky aluminum samples and resulted in larger cell sizes, probably as a result of the increased pressure by the blowing agent. Photomicrographs of foam sections from each sample revealed that the smaller, globular aluminum particles tended to be found at cell junctions (*9,14*) while the larger thin flakes (flaky aluminum) tended to coat the cell walls. A photomicrograph of the foam containing flaky aluminum is shown in Figure 7. The aluminum particles indeed form a coating on the cell walls leading to the metallic appearance of the foam that ultimately leads to an optical effect.

Microgravity Experiment

Experiments designed to form an aluminized foam in microgravity were conducted in Consort-series rocket flights using the same apparatus used for the polyurethane foam described above (*10,11*). The only difference was that flaky aluminum was used as an additive to the polyol premix. The polyol premix contained all components other than the diisocyanate reactant (i.e. polyol, tertiary amine catalyst, blowing agent, and surfactant). In both experiments with this configuration, system failures prevented the primary goal from being achieved. In the first flight experiment, a burst seal of the wrong thickness prevented the polyol premix and the diisocyanate from mixing. Only the polyol premix was in the reaction container, and hence it was expelled without reaction. Nevertheless, excellent mixing of the several components of the polyol premix was achieved as with the earlier foam experiment containing no aluminum. Pictures of the polyol mixture foam being expelled into the outer experiment chamber during the microgravity phase of the flight were recorded. They showed that the premix hanging, by surface tension, onto the screen of the funnel and, because of the presence of the blowing agent, desiring to expand. Thus the picture shows a somewhat spherical uncured foam forming, Figure 8. The mirrors adjacent to the exit funnel reveal a more complete view of the uncured

Figure 6. Photograph of cross-sections of polyurethane samples prepared in the laboratory with granular aluminum (left sample) and flaky aluminum (right sample).

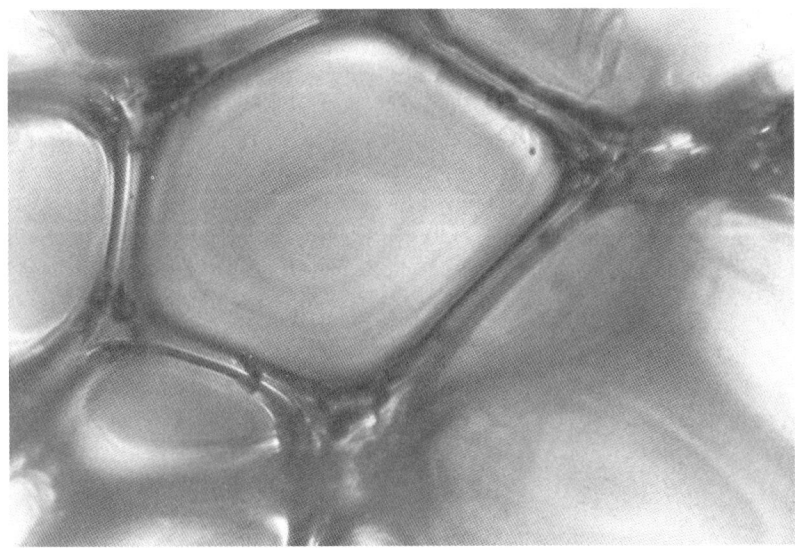

Figure 7. Photomicrograph (100X) of foam sample prepared with flaky aluminum. Cell diameter is ca. 0.5 mm.

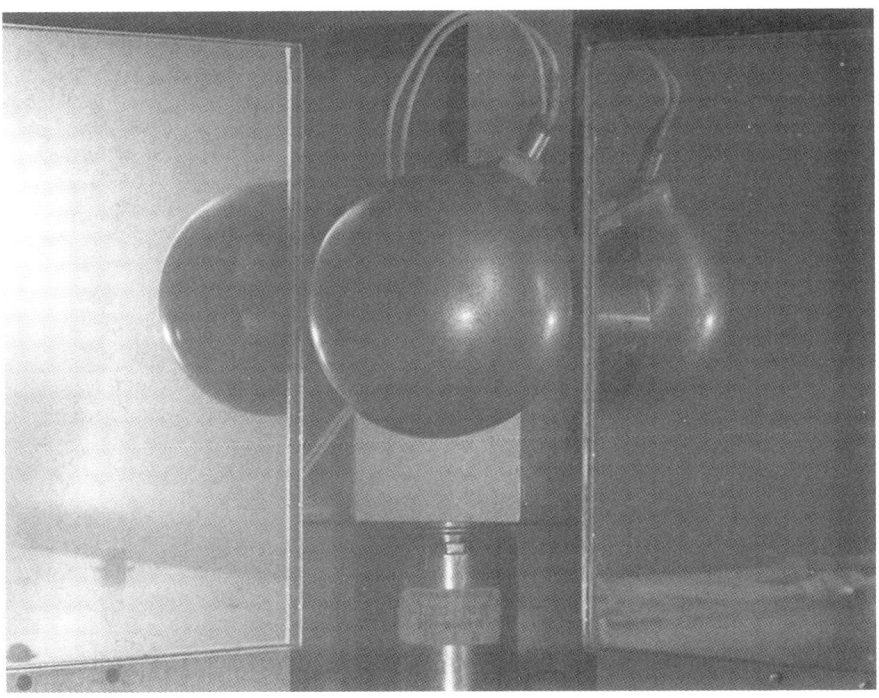

Figure 8. Picture of aluminized polyurethane premix being expelled from the mix chamber at ca. 1 atm. and at near zero g during a Consort rocket flight. Diameter is ca. 10 cm.

foam ball as it achieves a spherical shape. The metallic aluminum appearance is obvious from the photograph.

In a second experiment of the same configuration on another Consort rocket, the rocket control system failed and the rocket spun out of control never having achieved its period of microgravity. Nevertheless, the experimental apparatus performed flawlessly and a foam was formed. Unfortunately, the rocket was spinning and thus producing several Gs. The resulting foam was deposited on the inside walls of the payload section of the rocket. Samples of the pancaked polymer removed from the payload section after rocket recovery revealed uniform incorporation of the aluminum flakes and a well-cured polyurethane. As the polymer was not foam-like, no analysis of the cells was attainable. Nevertheless, we achieved sufficient knowledge about the uniformity of the foam to make a third experiment of this configuration unnecessary.

Effects of the Vacuum of Space (15)

Having succeeded in forming at least one high quality foam in a microgravity environment and having demonstrated that the microgravity environment imparted no negative effects on either the process or the foam properties, we set out to address the formation of a foam in vacuum. As anticipated, this task was challenging. Experiments run at partial vacuum levels proved that the normal foam formulations would not give an acceptable foam. Typically the blowing agent was stripped out leading to mostly open cell foam with distorted cells and with a structure that tended to be elongated toward the vacuum pump. Ultimately, to provide a desired shape for the foam and to partially overcome the loss of blowing agent, we employed a container such as a collapsed balloon or bag.

Attempts to use higher boiling point chlorofluorocarbon blowing agents did not satisfactorily solve the problems noted; in a high vacuum, the blowing agent would be stripped out of the foam leaving the resulting foam weakened and structurally inadequate. Ultimately, we demonstrated that water acceptably served in place of the blowing agent. Water reacts with the isocyanate leading to the production of carbon dioxide. Perhaps water acted both as a blowing agent and as a source of CO_2. The resulting foam tended to be spongy initially but slowly cured into a hard polymer foam. For example, when forming the foam into a beam in a vacuum in the laboratory, the beam appeared to have good qualities until the vacuum was released. Sample 2 (left picture) in Figure 9 is an example. Sample 6 (Right picture in Figure 9) shows a formulation that survives the atmosphere after release of the vacuum if sufficient curing time is allowed before release of the vacuum. Thus, optimum formulations were found that would lead to the formation of good foams for space structures. For the purpose

93

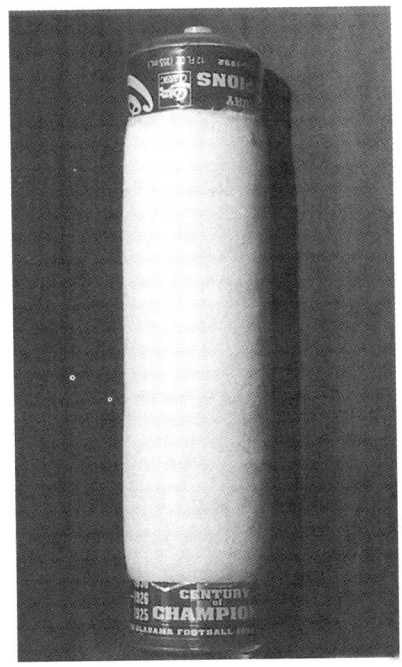

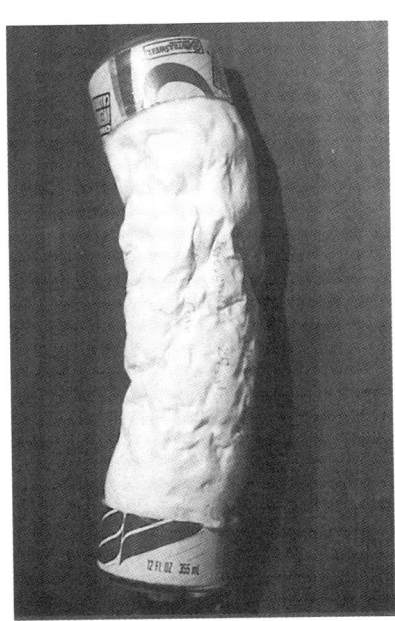

Figure 9. Photographs of foam beams prepared with different polymer mixes in vacuum in the laboratory.

of defining a shape, however, we ultimately used a container that employed a cloth cylinder (sock-like) that was capped with metallic ends (the ends of an aluminum soda can were used) through which foam premix could be added at one end, and pressure could be relieved, at the other end. Prior to the experiment, the cloth cylinder is stored within the metal caps giving a total pre-beam length of approximately 7 cm. The fully deployed beam achieved a length of approximately 35 cm in our configuration.

Using a foam formulation somewhat similar to that used in earlier rocket flights but with water replacing the blowing agent, a foam deployment system was constructed which allowed formation of foam beams in the vacuum of space. Simple beams were designed and constructed for use for foam containment. For example, discharging the mixed foam components into a cylindrical sleeve led to the formation of a long cylindrical beam. Two Consort-series rockets included foam experiments to demonstrate (1) foam formation in a round rubberized container and (2) foam formation in a cylindrical sleeve (see Figure 10). The payload section was vented to the atmosphere to achieve deployment of the foam in a vacuum. Both flight experiments were successful. The latter experiment demonstrated beam formation and the concept of pop-up foam structures. A video taken during foam deployment revealed the perfect operation of the hardware and perfect foam formation. In the latter experiment, a foam beam with an excellent structural appearance formed. The flight apparatus, after recovery, shows that the formulation used in this flight had an insufficient time to fully cure as some shrinkage occurred due to re-pressurization during re-entry just about 3 minutes after foam formation. In a space deployment, of course, the system would not be re-pressurized and a fully cured polymer would have resulted.

Conclusions

Polyurethane foams formed in microgravity have more nearly perfect cell structures. The cells are more regularly shaped and typically smaller. The foams formed in microgravity also have smaller voids. Addition of flaky aluminum particles to the foam recipe leads to a foam with a metallic appearance. Polyurethane foamed structures can be formed in space by replacing the normal blowing agent with water. The resulting structures provide the potential for the preparation of lightweight structures in space for use in space.

Acknowledgement

The authors gratefully acknowledge the efforts of the many individuals and groups that collaborated in this effort. The National Aeronautics and Space

Figure 10. Photograph of a foam beam formed during a Consort launch.

Administration provided financial support, flight opportunities, and technical support. Hercules, Inc., led by Debra J. Weiker, and Thiokol Corporation, led by Dean Lester, provided financial and technical support. The McDonnell-Douglas (now Boeing) integration team, headed by George Maybee, provided top quality launch support.

References

(1) Inquiries may be directed either to Prof. S. P. McManus, mcmanuss@uah.edu or Prof. F. J. Wessling, wesslif@uah.edu.
(2) Saunders, J. H.; Frisch, K. C. *Polyurethanes: Chemistry and Technology: Part I, Chemistry*. Wiley-Interscience, New York, **1962**, pp. 261-343.
(3) Bruins, P. F., ed. *Polyurethane Technology*, Wiley-Interscience, New York, **1969**, pp. 67-77.
(4) For a recent review, see Klarreich, E. G. *American Scientist*, **2000**, *88*, No. 2, 152. The structures in Figure 1 can be constructed by use of computer program of Prof. Ken Brakke, e.g. see www.susqu.edu/FacStaff/b/brakke/kelvin/kelvin.html.
(5) Naumann, R. J.; Mason, E. D. *Summaries of Early Materials Processing in Space*, NASA TM-78240, Aug. **1979**. More recent experiments that provide excellent data on microgravity effects with fluids can be found in the accompanying preprints from this symposium.
(6) e.g. See refs. cited in Noever; D. A.; Cronise, R. J.; Wessling, F. C.; McManus, S. P.; Matthews, J. H.; Patel, D. *J. Spacecraft and Rockets* **1996**, *33*, No. 2, 267.
(7) Murr, L. E.; Kinard, W. H. *American Scientist* **1993**, *81*, 152.
(8) Srinivasan, V.; Banks, B. A., eds. *Materials Degradation in Low Earth Orbit*, The Minerals and Mining Society, Warrendale, PA, **1990**.
(9) Unpublished results, McManus, S. P.; Wessling, F. C.; Patel, D.
(10) Wessling, F. C.; Maybee, G. W. *J. Spacecraft and Rockets* **1989**, *26*, No. 5, 343.
(11) McManus, S. P.; Wessling, F. C.; Matthews, J. H.; Patel, D.; Weiker, D. J. *Proceedings VIIth European Symposium on Materials and Fluids Sciences in Microgravity*, Oxford, UK, Sept. 1989, *ESA SP-295* (January, 1990).
(12) Wessling, F. C.; McManus, S. P.; Matthews, J. H.; Patel, D. *J. Spacecraft and Rockets* **1990**, *27*, No. 3, 324.
(13) Noever; D. A.; Cronise, R. J.; Wessling, F. C.; McManus, S. P.; Matthews, J. H.; Patel, D. *J. Spacecraft and Rockets* **1996**, *33*, No. 2, 267.
(14) Kanner, B.; Decker, T. G. *J. Cell. Plast.* **1969**, *5*, 32.
(15) McManus, S. P.; Wessling, F. C.; Emoto, K.; Patel, D.; Howard, D., unpublished results.

Chapter 7

Gel Formation under Microgravity Conditions

V. A. Briskman and K. G. Kostarev

Institute of Continuous Media Mechanics, RAS, Perm, Russia, Ak. Korolev Str., 1, Perm, 614013, Russia (fax: +7 3422 336957; email: vab@icmm.ru)

> This paper presents the analysis of the orbital experiment «Gel-1», performed on the «Mir» space station. The use of optical methods enabled us to visualize the photoinitiated reaction of gelation. The examination of obtained specimens has shown that their optical and mechanical properties are more homogeneous than in specimens obtained in earth conditions. To elucidate the role of different gravitational sensitivity mechanisms the data from orbital experiments are compared with the results of laboratory.

Introduction

In 1992 the experiment, called "Gel-1", was carried out on the "Mir" space station. The goal of the experiment was to study the evolution of frontal photopolymerization in conditions of real microgravity. "Gel-1" was a logical sequel to the orbital experimental studies jointly performed by V.Leont'ev (the Institute of Bioorganic Chemistry, Tashkent), A.Mashinsky (the Institute of Medical-Biological Problems, Moscow) and G.Nechitailo ("Energy" Space Corporation, Moscow) in 1980-1988 (*1-4*) on board "Salyut" and "Mir" orbital stations. While pursuing here the objective other than a review of the literature we will only mention that experiments on polymer production have been undertaken also on other space vehicles and aircrafts (*5-9*).

The formation of poly(acrylamide) gel (PAG) by photoinitiation was used by Soviet scientists as a base process for investigation. During this series of experiments the investigators found, that gels produced in laboratory conditions on the earth and on the orbital stations showed different properties. In particular, the matrices prepared from the orbital PAG specimens had much higher resolution in the case of electrophoresis of protein mixtures. This improvement of the polymer properties was supposed to be the result of weakness of gravitational sensitive mechanisms involved in gelation and essentially affecting its evolution. In this respect, the most contributing gravitational mechanism was evidently free convection developed in a liquid monomer during reaction. This hypothesis was supported by experiments and numerical calculations (*10-14*). In polymerization under laboratory conditions, convection caused by an exothermicity of the process and formation of a new, more dense phase stirs the reaction mixture, changing the distribution of the monomer conversion and giving rise to inhomogeneity of the final gel structure. By contrast, a reaction proceeding in weightlessness provides conditions for polymerization in a motionless monomer, which allows preparation of homogeneous PAG specimens.

"Gel-1" was expected to continue the previous series of experiments and to ensure the succession of experimental research by utilizing the reaction mixture formulated in (*1*) and applying the analogous source of initiating illumination. On the other hand, "Gel-1" was intended to serve as a launching-site for a new series of orbital experiments, aimed at gaining a comprehensive understanding of polymerization in microgravity. For this purpose a series of pioneering visualization and thermal measurements of the gelation process was made. The main goal was to validate the hypothesis on the frontal nature of photopolymerization and to derive equations of front propagation. Furthermore, this experiment could provide important information for designing structures of base reactors (cuvettes), and updating the methods of space experimental studies and laboratory analysis of the obtained specimens.

Experimental equipment

To investigate gel formation and to detect any possible convective flows in a mixture under orbital flight conditions the researchers used a shadow device "Pion-M", which provided synchronous photographs and temperature measurements of the examined object.

Two special cuvettes were manufactured for "Gel-1" experiment. The general view and schematic representation of the cuvette are given in Figure 1. A cuvette used for experiments was a rectangular cell formed by parallel-plane glass plates, 10 mm thick. The inside dimensions of the cell were $70 \times 30 \times 10$ mm. The wide facets of the cuvette, when properly adjusted, made a working cell of a Fizeau interferometer. This allowed one to thoroughly examine the optical structure of the obtained gel specimens, as soon as they were transported to the Earth. A glass cell was enclosed in an aluminum case serving as a means for adjusting and fastening the cuvette in the shadow device. The case had viewing windows on the side of the wide cell facets and the holes for reaction photoinitiation on the side of the long narrow facets.

An SD1-7 lamp was used a source of initiating radiation. It is made up of two gas-discharge lamps covered from inside with special fluorescent substance. The lamps are protected by a flat, light dispersing shade. The axial light intensity of the lamp block is 44 candela. The maximal illumination intensity of the lamps lies in the range of $\lambda = 4450$ Å (Figure 2), which coincides with one of the spectral response peaks of the photoinitiator. The degree of illumination provided by the CD1-7 at the cuvette inlet is *E =1200 lux.*

The cuvette was placed in a light-tight rigid case to exclude incidental photoinitiation of reaction during storage and transportation. When designing the case a provision was made for its removal after installation of the cuvette in the shadow device just before switching on the lamp.

To measure temperature of the exposed 1 (Figure 1) and opposite 3 facets the cuvette was fitted with two copper-constantan thermocouples 4. These measurements were made to define a temperature gradient caused by photopolymerization and by possible side effects - thermal radiation of the initiating source.

Figure 1.The general view (a) and schematic representation (b) of cuvette (viewed from the side of wide facet)

The cuvette was filled with the reaction mixture through the orifice on the upper facet. This orifice was also used to house a monomer dilatation compensator in the form of a rubber bell-jar 2.

A reaction mixture used in the experiments was composed of 18% water solution of acrylamide and small additives of crosslinking agents (methylenebisacrylamide, 0.46%), catalyst (tetramethylenediamide, $1 \cdot 10^{-2}$ %) and initiator (riboflavin, $4.9 \cdot 10^{-4}$ %). The latter is the main light absorbing component of the mixture. Several bubbles of argon were introduced in the bulk of the monomer to detect any convection.

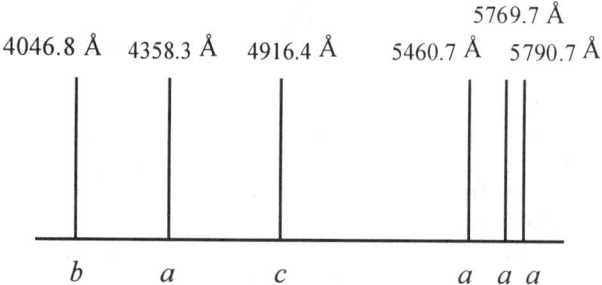

Figure 2. *The spectrum of SD-7 lamp, showing lines a, b, c having high, medium, and low intensity, respectively*

Terrestrial modeling and space experiment preparation

The orbital experiment was performed to study the forced frontal propagating polymerization. The frontal evolution of this reaction was caused by the light gradient arising due to the light absorption by a photoinitiator. The region, in which the light intensity decreased according to Beer's law, penetrated deep into the cuvette. The reaction front moved together with this region (*15*). In the described experiment, two characteristic parameters – the reaction front width and the distance along which the light was absorbed - were comparable with the cuvette length in the direction of the front propagation. This was most clearly observed in terrestrial experiments, in which it was possible to build the chemical reactor of larger length than in space.

To estimate the ability of "Pion-M" to detect gel inhomogeneities, visualization of polymerization in the same experimental cuvette under lateral light exposure was carried out in terrestrial conditions. As it is readily seen from the shadow picture (Figure 3a), a convective motion arising under the action of gravity is of a single-vortex type. A liquid monomer with a high conversion degree moves upwards along a thin polymer layer at the exposed facet. Then, while cooling and polymerizing further, it goes down in the form of separate jets. These jets being the center of polymerization create large conversion gradients detected by "Pion-M". These jets are also responsible for formation of essential inhomogeneities in the final polymer structure (Figure 3b).

Therefore one might expect that in microgravity conditions the marked optical inhomogeneities caused by a strong convection would be fixed with the "Pion-M" shadow device, whereas small inhomogeneities caused by a weak motion would be detected with an interferometer after delivering the cuvette back to Earth.

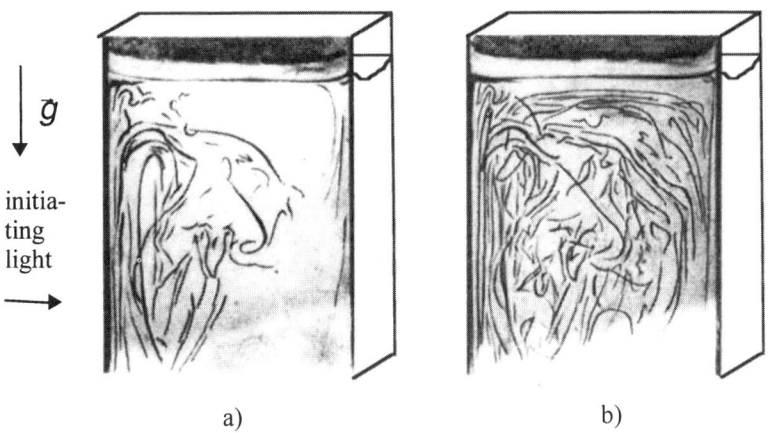

Figure 3. Shadow pictures of convective motion (a) and final structure of gel (b) formed in terrestrial conditions

It is known (*12, 16-18*) that the intensity of the convective motion is defined by the dimensionless Grashof parameter Gr involving, as a cofactor to a summarized acceleration of mass forces g_s, the characteristic temperature difference. When gelation proceeds in microgravity conditions, the level of g_s decreases but the heat change between cuvette and surrounding medium becomes worse. This leads to the increase in temperature gradients in the mixture during reaction, which in turn provides the conditions under which the convection may occur, even when $g_s = 10^{-4} g_o$. Therefore to prevent initiation of convection, the heating of the mixture was reduced by increasing the cuvette mass compared to that of a liquid. In this case the cuvette is a heat sink.

In order to estimate the effect of adiabatic conditions on the evolution of reaction, we performed a series of laboratory tests on polymerization in a cuvette placed into a foam-plastic case with organic glass windows for observation and photoinitiation. The reaction mixture - cuvette mass ratio was 1:20. The experiments showed that in this condition the process of gel formation accelerated. However, the increasing cuvette mass allowed us to reduce the maximal increase of the cuvette temperature to 10 °C (in the earlier series of orbital experiments the overheating measured several tens of degrees). This mass-to-mass ratio was kept unchanged in our orbital experiment.

There is one more aspect worthy of note. It is very important that by the time of running experiments on board space station (this could be a rather long period, about several months) the monomer mixture should retain its reactivity. To ensure this ability, the reaction mixture was degassed before filling the standard cuvettes. Degassing was by putting samples under a vacuum followed by bubbling the reaction solution with an argon. This procedure was used as an effective means to remove a molecular oxygen, which was the reaction inhibitor, (oxygen can react with polymer radical, forming less reactive peroxy radicals).

In addition, prior to reaction, the cuvettes with the monomer were kept at temperature of 10 °C.

Results of space experiment

After completion of the orbital experiment the movie film of the performed tests and cuvettes were transported to the Earth (*19-23*). The examination of the cuvette contents (Figure 4) showed that in both cases the process of gel formation involved photoinitiated polymerization of the reaction mixture.

However in a part of cuvette capacity was found an opaque substance consisting of interlinked white grains with the density higher and the properties other than those of a PAG gel.

Formation of opaque polymer ("popcorn" or "omega"-polymer) in gelation is of spontaneous character and is caused by a generation of free monomer radicals in the absence of photoinitiator participation. In the present experiment, this can be attributed to a non-fulfillment of proper storage conditions during transportation, in particular, to an increase of the monomer temperature to about 40 - 60°C.

Generation of opaque polymer reduces the initial concentration of the monomer. Further laboratory investigations have demonstrated that in one case (Figure 4a) this lowers the monomer concentration to 9.5% and in the other case - to 13% (Figure 4b). It is to be noted, that in both cuvettes the opaque polymer is localized in the same part of the cavity, which is an evidence of its formation in the conditions of existing gravitational field or under overloading. The analysis of shadow pictures of cuvettes at times prior to exposure has revealed that the polymer particles are surrounded by a layer of gel (Figure 5a), 1-2 mm thick. Evidently, this layer serves as a source of gel globules arising in a liquid monomer. The fact that the reaction mixture before photoinitiation was in a liquid state was supported by the spherical shape of the gas bubbles. The size of the bubbles ranges from several fractions of a millimeter to 5mm.

Figure 5 presents a series of cuvettes photographs at different times upon exposure to light (initiation from below). As it is readily seen from the photographs, the gas bubbles remain fixed during the reaction. This means that the gelation proceeds layer-by-layer, and the convection is absent, which is in correspondence with the small Grashof value not exceeding 30. However, the shadow device did not record the propagation of the reaction front which is expected to have the form of a system of parallel lines. The spontaneous polymerization reduced the monomer concentration in the reaction mixture and, therefore, the conversion gradient, directly in the PAG front, appeared to be smaller than the sensitivity threshold of a shadow device.

Although a direct observation of the front propagation was not feasible, the progress of reaction in the bulk of the monomer was followed with the aid of argon bubbles. Consider a bubble illuminated with a parallel light beam with the intensity I_0 (Figure 6). Since the refraction coefficient of a gas in a bubble is less than that of the surrounding liquid, the bubble in this case plays the role of a diverging lens. Figure 6 schematically represents the distribution of light intensity behind the bubble in the cross-section A-A. It is seen, that in the region a-b there is an overlapping of the two light fluxes - the initial flux with the intensity I_0 and the flux dispersed by the gas bubble. A drastic difference of intensity along the boundary of the bubble "shadow" gives rise to a corresponding difference of the polymerization degree. It is just this gradient that is recorded by the shadow device. Since the limiting sensitivity of the shadow device corresponds to a certain conversion gradient, one can use this conversion degree as a marker of front propagation with time (Figure 7). Following this figure we can suggest that in microgravity, as well as in terrestrial conditions, the variation in conversion of the reaction mixture does not occur at once, but rather after a certain period of time necessary to initiate the reaction (the induction period). The generation of opaque polymer in the bulk of monomer increases the photoinitiation time t_0. Thus, in laboratory conditions, for a similar initial monomer concentration the time for initiation was $t_0 \sim 4$-5 min, whereas in orbital conditions is $t_0 \sim 6$-7 min for cuvette N2 and is $t_0 \sim 7$-8 min for cuvette N1. Note, that the induction period is longer in a monomer with a greater content of opaque polymer.

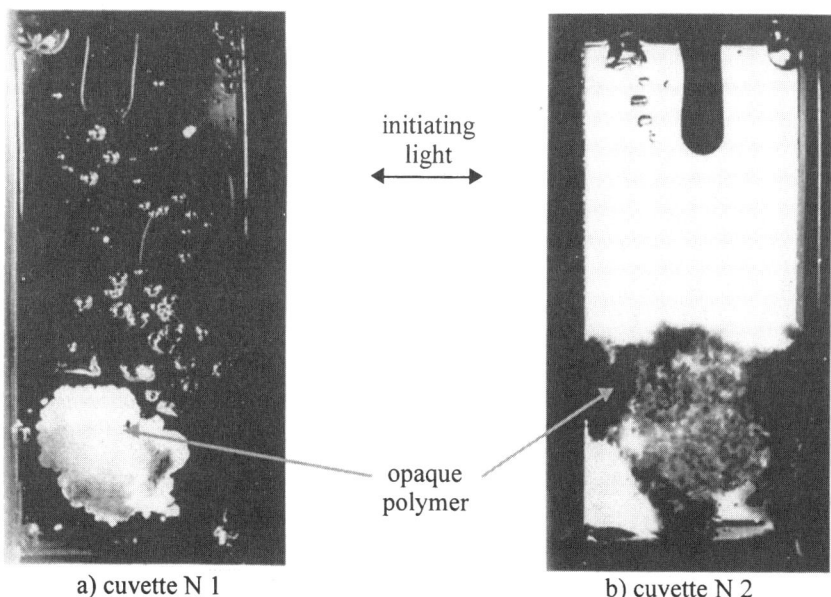

a) cuvette N 1   b) cuvette N 2

Figure 4. Photographs of gel specimens after return to the Earth. The photographs were made in reflected (a) and transmitted (b) light

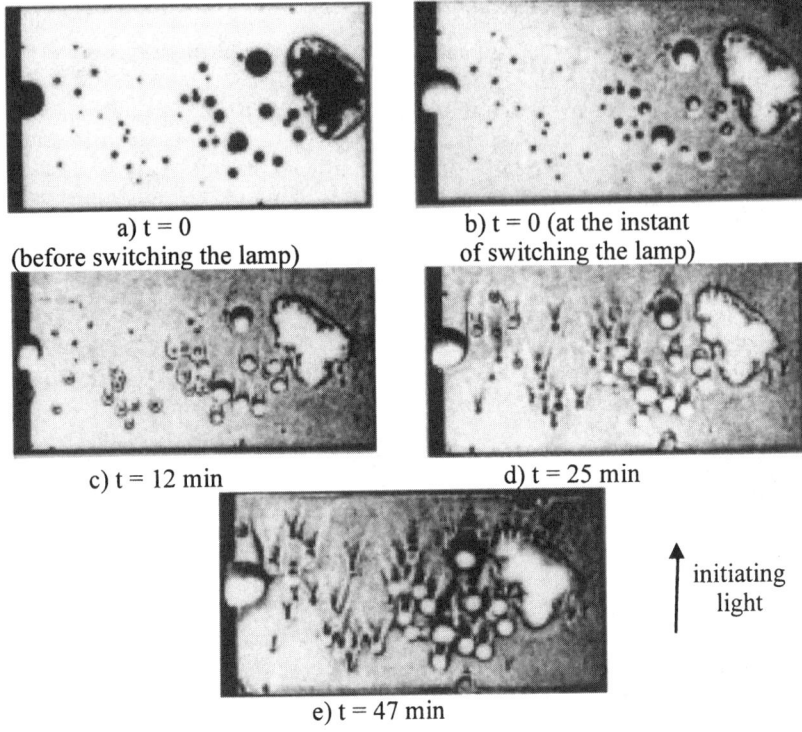

Figure 5. Photographs of reaction mixture during gel formation process in microgravity

It is evident from Figure 7, that within the experimental error the behavior of relationship d = f(t) is well described by the logarithmic law d/d_0 = A lnt + B, where d_0 is the cuvette dimension in the direction of initiation, and A and B are constants. This equation was proposed by the authors based on the phenomenological model of polymerization front propagation in the absence of convective heat/mass transfer. Similar equations were obtained analytically in (24). Solution of the analytical equation gives

$$A = \frac{2}{\varepsilon}; \quad B = \frac{2}{\varepsilon}\left[\ln(k_p(\varphi\varepsilon I_0 c_0 / k_0)^{1/2}) - \ln\ln\frac{M_0}{M}\right],$$

where ε is the coefficient of extinction, I_0 is the intensity of the incident light, φ is the quantum output of the initiation reaction, M_0 and c_0 are the initial concentrations of the monomer and initiator, M – concentration of the monomer at the front taken as a reference unit, and k_p and k_0 are the rate constants of the chain growth and termination (25). The value of ε, calculated from the slopes of the lines, is 0.62 cm^{-1}.

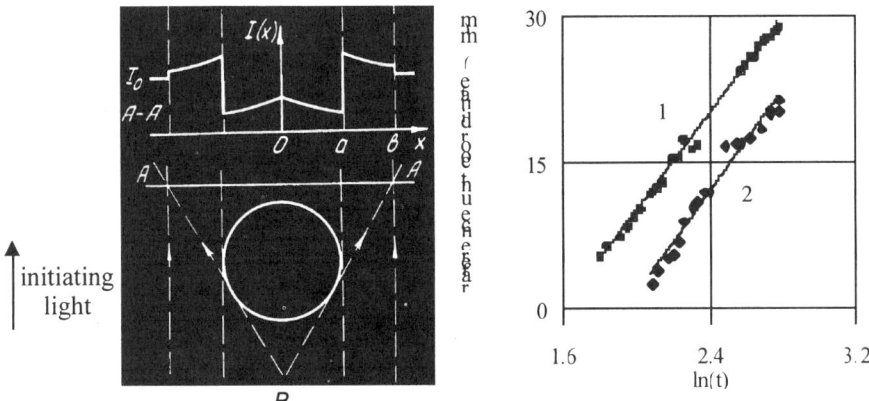

Figure 6. Distribution of light intensity behind the gas bubble in liquid across the A-A section, B – apparent focus of dispersing bubble- lens

Figure 7. Polymerization front position versus of the logarithm of time. Concentration of acrylamide: 1 – 13%, 2- 9.5%

It should be noted that, the maximum conversion gradient observed during the front propagation on the boundary of the bubble "shadow", shifts toward the dispersion boundary with the increase of the polymerization degree (Figure 6). Deformation of the bubbles occurs later: they have grown larger and elongated in the direction of initiating illumination due to a nonuniform monomer shrinkage. The results of such deformation are displayed in Figure 8.

More predictable are the results of thermal measurements. The temperature curves (Figure 9) have the same form as the curves obtained in laboratory experiments under adiabatic conditions. An adequate heat contact of the cuvettes and the shadow device and an essentially decreased monomer concentration lower the temperature increase of the cuvette to 1.5 °C. The effect of already completed spontaneous polymerization also allows for the fact that the temperature maximum moves to the region of large times (under earth adiabatic conditions the temperature maximum is reached at t = 25 min).

As mentioned above, the design of the standard cuvettes allowed us to use them as the working cells of a Fizeau interferometer. However, imperfection of fabricated cuvette cases made it impossible to adjust the cuvettes to a band of infinite width. The adopted adjustment procedure produced an optical wedge of 10 bands for cuvette N2 and of 150 band for cuvette N1. The bands were parallel to the illuminated facet.

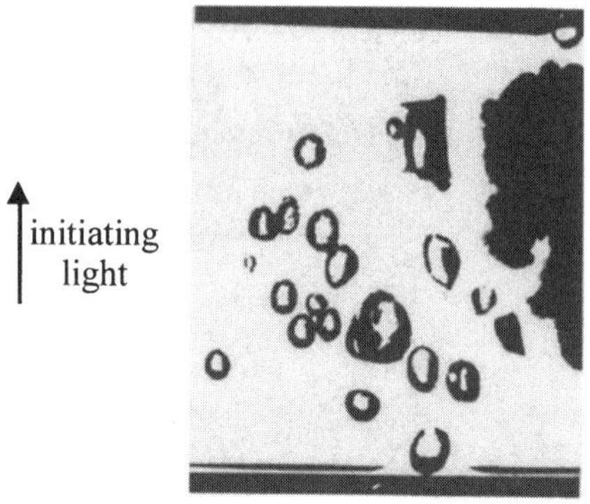

Figure 8. Bubble deformation in gel specimen in the central part of the cuvette N1 (at the scale of 1 to 0.75)

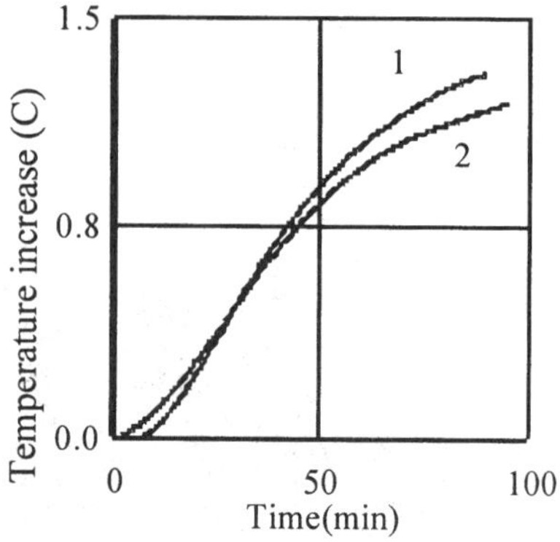

Figure 9. Temperature of the exposed (1) and opposite (2) faces of the cuvette N1

During the experiment both, the number and orientation of interference bands were changed. Figure 10 shows an interferogram of gel occupied the central part of the cuvette N2 with the upper facet exposed to light. Here the opaque polymer is on the left, and the gel globules are on the right. The number of bands in the direction of the initiating illumination does not increase compared to a check interferogram made prior to the flight. This means that gel generated in microgravity conditions is much more homogeneous, than the laboratory specimens. The inclination of the bands has changed due to the effect of the omega-polymer (Figure 11).

During light exposure the omega-polymer, the gel globules and gas bubbles not only absorb and transmit the light, but also reflect it. The reflected light enhances the intensity of initiating radiation ahead of the scattering object and thus provide favorable conditions in order that gelation, occur before the arrival of the overall reaction front. Due to its earlier development, the gel in this region reaches a higher degree of conversion at the time of photoinitiation. The resulting inhomogeneities of the conversion field in the form of concentric bands are seen on the photographs (Figures 10 and 11).

An opposite situation arises in the region of the objects' shadows. Polymerization proceeds there at a lower rate than in the bulk of the monomer, and therefore the conversion degree is slightly less than the averaged one (on the photographs these zones are in the form of characteristic lenses (Figure 11).

Along with studying the optical structures of the orbital specimens, the other object of laboratory experiments was to investigate the distribution of

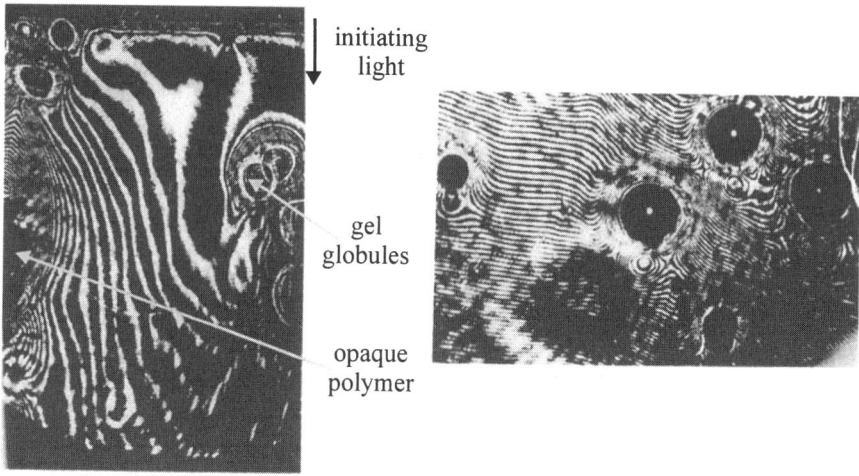

Figure 10. Interferogram of conversion field inhomogeneities caused by formation of opaque polymer and gel globules in the orbital gel specimen (at the scale of 1 to 0.35). Cuvette N2.

Figure 11. Interferogram of conversion field inhomogeneities caused by gas bubbles (at the scale of 1 to 0.24). Cuvette N1.

physico-mechanical properties of the gel. The studies were carried out on a installation for thermo-mechanical tests of polymers (ITMT-70). As expected, the orbital specimen was more homogeneous than its laboratory analogs.

Conclusion

From the analysis of the obtained results one can draw the following conclusions (26):

The «Gel-1» orbital experiment fulfilled all the tasks put forward in spite of occurrence of spontaneous polymerization. It was proved that the real level of mass forces on the orbital station was not sufficient for development of convection during gel formation. Therefore the reaction of gel formation was of a frontal character (27 - 29). The optical methods were first used to investigate polymerization directly in microgravity. The time dependence of frontal localization during gel formation was established. The optical structure of the orbital specimens was thoroughly examined to define the reasons for formation of inhomogeneities and their character. The measurements of physico-mechanical properties demonstrated a higher homogeneity of the orbital specimens in contrast to their earth analogs.

The orbital experiment allowed the investigators to reveal the deficiencies of the cuvette design and of the experimental method, to improve the conditions for transportation and storage of the reaction mixture, and to develop new and updated procedures for laboratory processing of the obtained data.

Acknowledgment

This Chapter is dedicated to the memory of the authors' dear friends and colleagues V. Leont'ev and A. Mashinsky who initiated and took an active part in the studies on polymerization in microgravity.

The authors take the opportunity to express their gratitude to V. Levtov, V. Romanov, S. Gazin, T. Lyubimova and Russian Astronauts S. Avdeev, A. Serebrov and A. Solovyev for the useful and pleasant collaboration in preparing and performing space experiments.

The present work was supported partially by RFBR under Grants 94-01-01730 and NASA-RSA Contract under Project STAC SBT-12. The presentation of paper at American Chemical Society Spring Meeting-2000 was supported by the Petroleum Research Fund, administered by the ACS and by the USRA.

Literature Cited

1. Sadykov, A.; Leont'ev, V.; Mangutova, Y.; Nechitailo G.; Mashinsky A. *Dok.l Akad. Nauk. SSSR*, **1988**, *300*, 1004-1007.
2. Abdurakhmanov, Sh.; L.Bogatyreva, L; Briskman, V.; Levkovich, M.; Leont'ev, V.; Mashinsky, A.; Nechitailo, G. On Polyacrylamide Gel Formation by Photoinitiation under Terrestrial and Orbital Conditions. In: *Numerical and Experimental Modelling of Hydrodynamic Phenomena under Weightlessness*; Briskman, V., Ed.; Ural Branch RAS: Sverdlovsk, Russia, 1988, 120-128
3. Leont'ev, V.; Abdurakhmanov, Sh.; Levkovich M. In *Proc. AIAA/IKI Microgravity Science Symposium*, 1991, 273-277.
4. Mashinsky, A.; Nechitailo, G. *Space Biology. Studies at Orbital Stations;* Mir: Moscow, Russia, 1993; 560 p.
5. McManus, S.; Wessling, F.; Matthews, J.; Patel D.; Weiker, D. In *Proc. VIIth European Symposium on Materials and Fluid Sciences in Microgravity;* Kaldeich, B., Ed.; ESA SP295; ESA Publications Division: Oxford, UK, 1989, pp 519-524.
6. Avci, D.; Thigpen, K.; Mathias, L. *Polym. Prepr.(Am.Chem.Soc.Div.)* **1996**, *37*, 317.
7. 1996-97 Biennial Report Consortium for Materials Development in Space; NAGW-812; University of Alabama in Huntsville: Huntsville, ALa, 1997

8. Frazier, D.; Paley, M.; Penn, B.; Abdeldayem H.; Smith, D. Microgravity Processing and Photonic Applications of Organic and Polymeric Materials. In *MSFC Center Director's Discretionary Fund Final Report, Project No.95-26,* NASA Technical Memorandum 108533, 1997.
9. Pojman, J.; Volpert, V.; Dumont, T.; Ainsworth, W.; Chekanov, Y.; Masere, J.; Wilke, H. Bubble Behavior and Convection in Frontal Polymerization on the KC-135 Aircraft, *AIAA 2000-0856, 38th Aerospace Sciences Meeting,* 2000.
10. Bogatyreva, L.; Briskman, V.; Levkovich, M.; Leont'ev, V.; Lyubimova, T.; Nechitailo, G. *Space science and technology.* **1989**, *4*, 43-47
11. Abdurakhmanov, Sh.; Babskii, V.; Briskman, V.; Leont'ev, V.; Levkovich; Mashinsky, A. In: *Proc. Gagarin Scien. Readings on Astronautics and Aeronautics,* Avduyevsky, V., Ed.; Nauka: Moscow, Russia, 1990, 219-228
12. Bogatyreva, L.; Briskman, V.; Kostarev, K.; Leont'ev, V.; Levkovich, M.; Lyubimova, T.; Mashinsky, A.; Nechitailo, G.; Righetti, P. In *Proc. VIII Europ. Symp. on Material and Fluid Sciences in Microgravity;* Kaldeich, B., Ed.; ESA SP-333 Vol.1; ESA Publications Division: Brussels, Belgium, 1992, pp 173-178.
13. Pojman, J.; Craven, R.; Khan, A.; West, W., *J. Phys. Chem.* **1992**, *96*, 7466-7472.
14. Bowden, G.; Garbey, M.; Ilyashenko, V.; Pojman, J.; Solovyov, S.; Taik, A.; Volpert, V. *J. Physical Chemistry* **1997**, *101*, 678-686.
15. Grishchenko, V.; Maslyuk, A.; Gudzera S. *Liquid Photopolymerizable Mixtures;* Naukova Dumka: Kiev, Ukraina, 1985; 206 P. (In Russian)
16. Golubev, V.; Korolev, B.; Kostarev, K.; Lyubimova, T. *Heat Transfer Research* **1993**, *25*, 888-893.
17. Malkin, A.; Begishev, V.; Guseva, L.; Kostarev, K. *Polymer Sciences* **1994**, Ser.A, *36*, 625-630.
18. Briskman, V.; Kostarev, K.; Yudina, T. Comparative investigations of thermal and photopolymerization under action of centrifugal forces. Basic mechanisms of heat/mass transfer. In *Centrifugal Materials Processing.* Regel, L.L.; Wilcox, W.R., Eds.; Plenum Press: N.Y., 1997, pp 247-255.
19. Briskman, V.; Kostarev, K. In *Proc. Int. Aerospace Congress;* Liberzon, M.; Ed.; STC "Petrovka": Moscow, Russia, 1995, Vol.1, pp 514-525.
20. Briskman, V.; Kostarev, K.; Levtov, V.; Lyubimova T.; Mashinsky, A.; Romanov, V.; Nechitailo, G. *Acta Astronautica* **1996**, *39*, 395-402.
21. Briskman,V.; Kostarev, K.; Levtov, V.; Romanov,V.; Yudina, T. Comparative experimental research of polymerization on the "MIR" station and on the Earth, *AIAA 95-0263 33rd Aerospace Sciences Meeting,* 1995.
22. Briskman, V.; Kostarev, K.; Levtov, V.; Mashinsky, A.; Nechitailo, G. Romanov, V. On gravity dependence of polymerization, *AIAA 96-0257. 34th Aerospace Sciences Meeting,* 1996.
23. Briskman, A.; Kostarev, K.; Matveenko, V.; Shardakov, I.; Yudina, T. Pistzov, N. In *Proc. Joint Xth European and VIth Russian Symposium on*

Physical Sciences in Microgravity, Avduyevky, V., Ed.; IPM RAS: St.Peterburg, Russia, 1997, pp 244-247.
24. Ivanov, V.; Smirnov, B. *Vysokomolekularnie soedineniya (High molecular compounds)*, Short Reports. **1991**, *33*, 807-810. (In Russian)
25. Ivanov, V.; Begishev, V.; Guseva, L.; Kostarev, K. *Polymer Sciences* **1995**, Ser.B, *37*, 293-296
26. Briskman, V., *Advanced Space Research* **1999**, *24*, 1199-1210.
27. Briskman, V.; Kondyurin, A.; Kostarev, K.; Leont'ev, V.; Levkovich, M.; Mashinsky, A.; Nechitailo, G.;Yudina, T. Polymerization in microgravity as a new process in space technology. *IAA-97-IAA* 12.1.07 *48th International Astronautical Congress,* 1997.
28. Kostarev, K.; Yudina, T.; Lysenko, S. *Polymer Sciences* **1998**, Ser.B, *40*, 386-390.
29. Briskman, V.; Kostarev, K.; Moshev, V.; Guseva, L. *Int. Polymer Science and Technology* **1998**, *25*, 62-66.

Parabolic Flight Investigations

Chapter 8

Bubble Behavior in Frontal Polymerization: Results from KC–135 Parabolic Flights

William J. Ainsworth, John A. Pojman*, Yuri A. Chekanov, and Jonathan Masere

Department of Chemistry and Biochemistry, University of Southern Mississippi, Hattiesburg, MS 39406

Frontal polymerization is a mode of converting monomer into polymer via a localized exothermic reaction zone that propagates through the coupling of thermal diffusion and the Arrhenius reaction kinetics of an exothermic polymerization. Studies were carried out aboard the NASA KC-135 aircraft to determine the effects of gravity on the interactions of bubbles with descending polymerization fronts with monofunctional and difunctional acrylates. The absence of buoyancy allowed bubbles to grow larger ahead of fronts of diacrylate polymerization. Bubbles formed in thermoplastic fronts appeared to aggregate and form periodic patterns behind the fronts but the poor quality and short duration of the low gravity precludes definitive conclusions about the mechanism.

Polymeric foams are a very important class of materials. They serve a wide variety of purposes from the insulation of refrigerators, the packaging of foods, and padding of chairs to the construction of lightweight but durable building materials. (*1*) Commercial production of polymeric foams was first introduced more than fifty years ago, and since that time much work has been carried out in an effort to improve the current production methods in order to produce foams with specific properties suitable for specific environments. (*2*) With the emergence of new technologies and production techniques, it is now possible to produce a polymer foam using almost any polymeric material. (*3*)

Wessling et al. reported the first polyurethane foam produced in reduced gravity in which they found major differences in the foam and cell structures formed, compared to those formed on Earth. (*4,5*) Bergman proposed that the space-based production of foams would prevent imperfections in the finished product caused by cell drainage and sedimentation. (*6*) Curtin et al. sought to produce high quality foams in reduced gravity, free of defects in order to test heat transfer theories and for uses as a standard reference material. (*7*)

Studies have been carried out under reduced gravity to determine the effects of gravity on the properties of foams produced through other methods. (*5,7*) These experiments found that gravity can have significant effects on the size and shape of cells formed, the density of the foam produced, and distribution of the cells within the polymer.

We sought to develop a relatively simple method for the production of polymeric foams that does not require mixing or continuous heating. Frontal polymerization offers this possibility. We undertook the first stages of developing this technology using the simplest systems possible. Because no surfactant or nucleating agents were employed, large bubbles formed, unlike commercial foams. However, the systems used provide information on how bubbles affect frontal polymerization and are affected by it.

Frontal Polymerization

Frontal polymerization is a mode of converting monomer into polymer via a localized exothermic reaction zone that propagates through the coupling of thermal diffusion and Arrhenius reaction kinetics of an exothermic polymerization. Frontal polymerization was discovered in Russia by Chechilo and Enikolopyan in 1972. (*8*) The literature up to 1984 was reviewed by Davtyan et al. (*9*)

Pojman and his co-workers demonstrated the feasibility of traveling fronts in solutions of thermal free-radical initiators in a variety of neat monomers at ambient pressure using liquid monomers with high boiling points (*10-12*) and with a solid monomer, acrylamide. (*13,14*) Fronts in solution have also been developed. (*15*) The macrokinetics and dynamics of frontal polymerization have been examined in detail (*16*) and applications for materials synthesis considered. (*17*) A patented-process has been developed for producing functionally-gradient materials. (*18*)

Frontal polymerization reactions are relatively easy to perform. In the simplest case, a test tube is filled with the initial reactants. The front is started by applying heat to one end of the tube with an electric heater. The position of the front is obvious because of the difference in the optical properties of polymer and monomer. The velocity can be affected by the initiator type and concentration but is on the order of a cm/min.

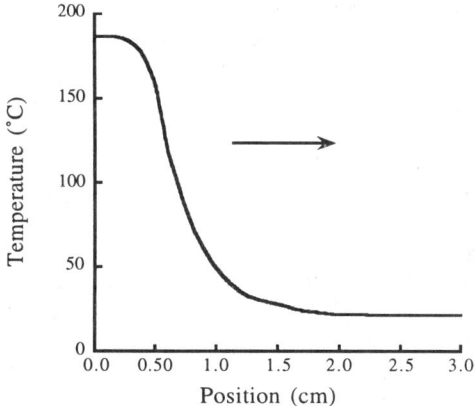

Figure 1. The temperature profile for a benzyl acrylate front, with ultrafine silica gel added to prevent convection.

The defining features of frontal polymerization are the sharp temperature and concentration gradients present in the front. Figure 1 shows a temperature profile for benzyl acrylate polymerization. Notice that the temperature jumps about 100 °C over 1 cm. The concentration gradient is two orders of magnitude larger.

Frontal polymerization can be achieved with a variety of monomers and has been studied with thermosets and thermoplastics. Examples include n-butyl acrylate, benzyl acrylate, styrene, dodecyl acrylate and hexyl acrylate. If the front is ascending, the monomer immediately above the front is lower in density because of the temperature gradient than the bulk monomer and so simple convection can occur for a thermoset (*19*) or a thermoplastic. (*20*) A descending front with a thermoset is stable but a thermoplastic is unstable because even though the polymer is very hot, it is more dense than the unreacted monomer. This leads to the Rayleigh-Taylor instability. (*16,21*)

Bubble interactions in frontal polymerization reactions were studied by Pojman et al. on the *Conquest I* sounding rocket with n-butyl acrylate (*21*) and were followed by a series of KC-135 parabolic flights studying thermosets. (*22-24*) They reported evidence of unusual bubble interactions in reduced gravity. In the *Conquest I* sample, they observed a periodic pattern of bubbles "connected like a necklace" with a region of polymer followed by a large bubble, followed by another region of polymer.

Because the video failed on the rocket flight, it was not possible to determine if the periodic structure resulted from a periodic mode of propagation or because of a thermodynamically-driven aggregation after the front. The periodicity could have arisen from a large bubble pinching off to minimize the free energy. However, there is precedent in frontal polymerization for periodic propagation modes. The expansion of poly(methacrylic acid) in a front, caused by gas from the initiator, was found to cause a periodic mode of propagation. (25)

Experiments performed aboard the KC-135 were inconclusive with benzyl acrylate (a thermoplastic) because of the short time period in weightlessness and the lack of pressure control in the reactor. Interesting results were obtained with triethylene glycol dimethacrylate (a thermoset) in which bubbles were produced by the peroxide initiator. (22-24) In high g, they found that long chains of bubbles would form and proposed that existing bubbles would act as nucleation sites for further bubble formation. In low g, the bubbles would move toward the center of the test tube through surface-tension induced-convection bringing the bubbles into closer contact where they could coalesce and grow. The pressure was not controlled during the reaction, and the velocity could not be independently varied from the amount of gas evolved because the initiator controlled the velocity and produced the gas. Therefore, we undertook a study in which a blowing agent was added to systems using a gas-free initiator and in reactors with a piston that allowed isobaric conditions.

The addition of an aqueous blowing agent to the reaction mixture results in the boiling of the blowing agent at the front. At the front, a viscosity gradient exists where the change from low conversion to higher conversion provides a soft "gel region" in which bubbles can nucleate and grow until the front passes, and the polymer becomes rigid. Bubble size can be controlled by varying the amount of blowing agent added, and the boiling temperature of the blowing agent. Bubble distribution can be controlled by varying the concentration of surfactant in the blowing agent mixture.

Bubbles are formed at the front where there is a sharp concentration gradient. The polymer in this region is soft and the bubbles can expand until the front passes, and the polymer gels and traps the bubbles. Figure 2 shows a frontal polymerization reaction in which bubbles are formed near the front and trapped in the polymer

In order to investigate the effects of gravity on the properties of foams produced via frontal polymerization, experiments were performed aboard the NASA KC-135 aircraft. Using frontal polymerization for reduced gravity studies offers both advantages and disadvantages compared to the previous reduced gravity experiments. (4,5,7) With multifunctional acrylates, the polymer gels rapidly so that the cells formed are not affected by the changes in gravity produced by the KC-135. This also allows us to view the effects of different levels of acceleration on the foam produced because the front propagates through several parabolas, which provides an internal standard for comparison of the effects of various levels of gravity. Another advantage is the simplicity -- no chemicals need to be mixed during flight.

Figure 2. A foaming system consisting of a monomer mixture containing HDDA, HEMA, Aliquat Persulfate, Aliquat 336, and Water. The piston on the bottom allowed the reaction to occur at constant pressure.

The main disadvantage is that the bubbles produced in the microgravity region of the parabola tend to coalesce and grow larger in an attempt to minimize their surface free energy. The bubbles become larger than the gel region of the front, and they are essentially pushed ahead of the front until the high g pullout. Another disadvantage is the elongation of the cells formed as they interact with the glass walls of the test tube. This allows for the overlap of the cells produced in one region of the parabola into the next region of the parabola. Also of major concern is the amount of time required to initiate the reaction, and the amount of time required for the front to propagate the length of the test tube. Systems with higher front velocities can be used but since the KC-135 can only produce 20 seconds of reduced gravity, the timing of the initiation of the reaction is very important in the acquisition of useable results.

Of course, for thermoplastics the polymer has a sufficiently low viscosity that during the high g phase, the bubbles rise and the polymer sinks.

Experimental

Liquid/Solid system

The reaction systems consisted of a monomer mixture containing 90% 1,6-hexanedioldiacrylate (HDDA; technical grade; Aldrich), with 10% 2-hydroxyethylmethacrylate (HEMA; Polysciences), which was added to slow the front velocity so that multiple parabolas could be seen in the same sample. A

10% Aliquat persulfate (tricaprylmethyl ammonium persulfate, APS), synthesized according to the procedure of Chekanov et al. (26) solution was added to the monomer mixture as a bubble-free initiator. This reaction mixture results in a rigid crosslinked polymer that will retain the cellular structure throughout the flight. The blowing agent solution was made up of 5% Aliquat 336 (Aldrich) and 95% distilled water. The Aliquat was added as a surfactant to allow the water to be dispersed in the organic monomer solution. All reagents were used as received.

The apparatus was the same as used on the *Conquest I* and described previously. (21) The reaction mixture was placed in a test tube, and a piston was inserted to allow for the expansion of the gas without a buildup of pressure inside of the test tube. A heater was attached, and the test tube was placed inside of a hermetically sealed glass chamber. The apparatus used is shown in Figure 3. When the target parabolas were achieved, the heaters were turned on and the reactions video taped.

Liquid/Liquid System

The reaction system consisted of 91% hexyl acrylate (Aldrich) as the monomer, 5% Aliquat persulfate as the free-radical initiator, and 4% Luperox 231 (1,1-di(t-butylperoxy)-3,3,5-trimethylcyclohexane; Atochem) as both an initiator and blowing agent because it decomposes to form free radicals and volatile products (methane and acetone).

The apparatus was built by the Marshall Space Flight Center (MSFC). It provided three levels of containment for two samples per flight, rapid ignition of the fronts and video. A three-axis accelerometer measured acceleration levels to 0.001 g, and the values can be seen in the images. A negative value indicates "normal" acceleration downward. For example, during the high g pull-out-, the middle accelerometer (2) would read -1.6 g.

Viscous Liquid/Liquid System

The reaction system was made more viscous by adding poly(hexyl acrylate) to the mixture. The poly(hexyl acrylate) was prepared by bulk polymerizing a 1% Luperox 231/ 99% hexyl acrylate solution in an oil bath at 110 °C for 30 minutes. The reaction mixture consisted of 20% of the polymer, 5% Aliquat persulfate, 2% Luperox 231, and 73% hexyl acrylate.

Apparatus

Diagrams of the apparatus used for the KC-135 flights are shown in Figure 3 below. The test tube containing the initial reactants and a piston to allow for the expansion of the gas was inserted into the glass containment chamber and a

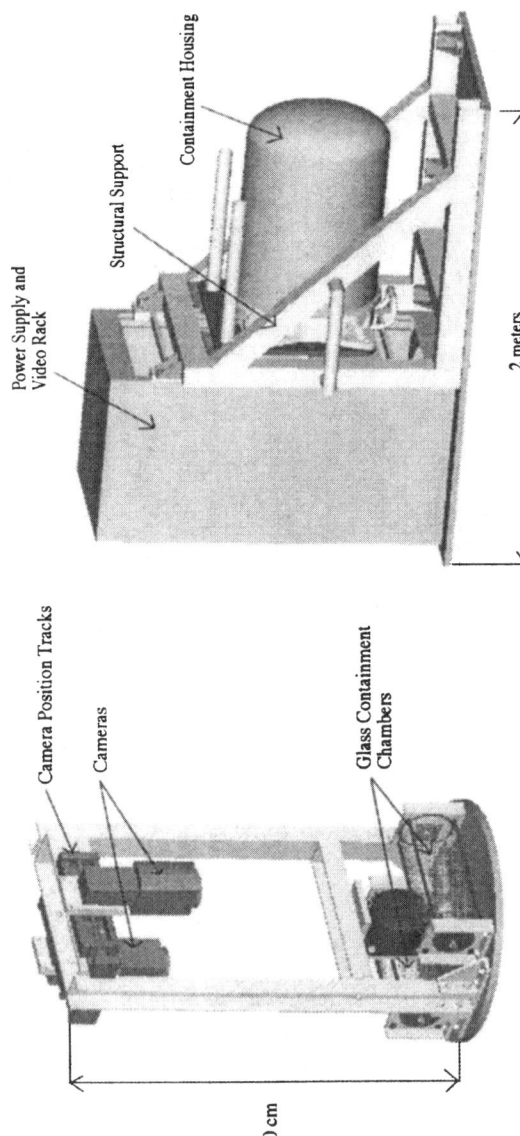

Figure 3. Diagrams of the apparatus used aboard the KC-135 aircraft for the liquid/liquid experiments.

cartridge heater was attached. Illumination of the reaction was provided by a flat panel light positioned behind the chamber and the reactions were monitored with video cameras mounted on adjustable positioning tracks. The apparatus was contained inside of an outer casing that provided a third level of containment and protected the internal components of the reactor. The entire apparatus is attached to the structural support and connected to the power supply and video control rack. Video images were recorded using digital VCR's on the rack.

Discussion

Liquid/Solid System

In laboratory-based experiments, we found that the size and distribution of the bubbles were nearly uniform (Figure 2). The bubble size increased as g decreased, but there was an overlap of the bubbles produced in microgravity into the region produced in high g as seen in the diagram in Figure 4. This overlap of phases of g-level eliminated our ability to measure the density of the foams produced, but the dramatic differences bubble size in each level leads us to believe that there would be a significant change in the density of the foam regions formed in microgravity compared to those regions formed in high g if samples were formed exclusively in low g. Because the systems propagated so slowly, this was not possible to achieve. A pure HDDA system would propagate rapidly enough but it was impossible to time the start of the front to overlap with the low g period.

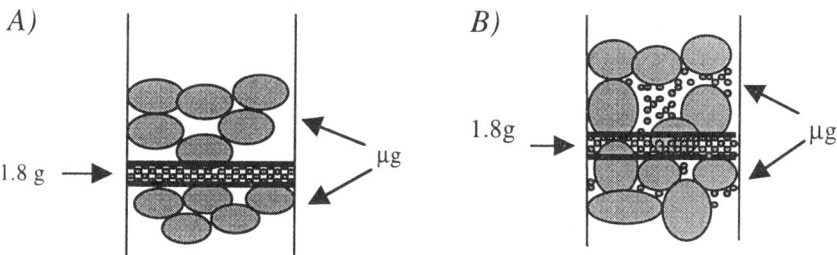

Figure 4. Schematics of expected (A) and observed (B) bubble size distribution.

Inspection of the video tapes and the samples from the flights also revealed one other interesting result. We found that in microgravity, the large bubbles were coalesced ahead of the front, but were surrounded by series of small cells when the high gravity phase was achieved. This can be seen in the images in Figure 5.

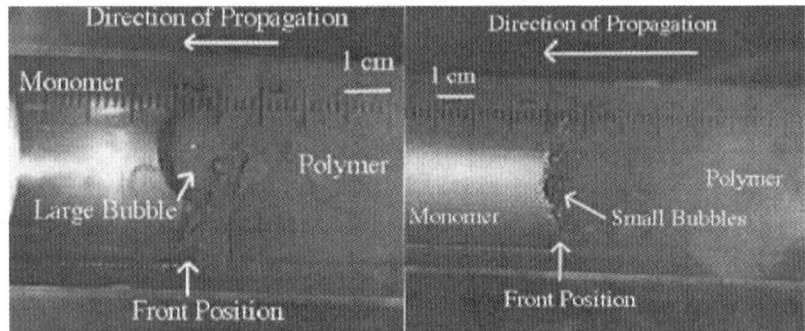

Figure 5. Images of foams produced in Low gravity (left) and High gravity (right). The systems consisted of HDDA, HEMA, Aliquat Persulfate, Aliquat 336, and water. The 1.8 g and µg images were taken from the same sample at different times during the flights.

Liquid/Liquid Systems

The use of a monofunctional acrylate provides a thermoplastic system that has the potential for use as an insulation and/or a cushioning foam. Since there is no crosslinking, the polymer does not become rigid but becomes viscous as it cools. In a system of this sort, the bubbles produced at the front are carried up by buoyancy. In weightlessness, there is no buoyancy, so the bubbles formed remain stationary, which allows the polymer to cool and increase in its viscosity while the bubbles are not moving. We found that aboard the KC-135, g-jitter causes the bubbles to move constantly. (Variations as large as ± 0.1 g were observed.) Several other dynamical variables also played a role in the problems observed. Thermocapillary migration of the bubbles toward the center of the test tube allowed many of the bubbles to coalesce to form very large bubbles in the front.

Bubble nucleation occurs at the front as the Luperox is decomposed. The bubbles begin slowly migrating upwards while they still continue to expand. The larger the bubble gets, the faster is the migration rate. Because of the piston, a void forms above the front. Another complication is that because of the high g period, the Rayleigh-Taylor instability caused polymer to sink, increasing the viscosity at the bottom of the tube.

We also observed migration of the bubbles toward the center of the tube as they rise. There are two possible mechanisms for this migration. The central axis of the test tube has a significantly higher temperature than at the walls. This causes thermocapillary migration of the bubbles toward the hotter region. (27) The second possible reason for the lateral migration is that convection behind the front causes the bubbles to be pushed by the convective current into the center. We have seen instances when a current can be seen to push the bubbles behind the front. On several occasions, we have observed bubbles near the walls being pushed downward as they grow, and when the size of the bubble becomes large enough, buoyancy carries the bubbles upward. In reduced gravity, the convection stops, and we observe the thermocapillary migration of the bubbles toward the center of the test tube where the temperature is the highest. Images taken from the KC-135 videos can be seen in Figure 6.

When the bubbles migrate toward the center, three phenomena occur. First, the bubbles become spherical because there is no buoyancy. The second is that fewer small bubbles are formed because the bubbles are essentially at rest, and most of the evolved gasses will feed into pre-existing bubbles. The third phenomenon is that as the bubbles grow larger, they begin to pinch off. (A drop can only be stretched to π times its diameter before it breaks. (28)) This offers a validation of the results obtained on the *Conquest I* sounding rocket. (21)

Increasing the Luperox concentration increased the size and number of bubbles but did not alter the qualitative behavior. Using more Aliquat persulfate increased the front velocity but did not affect the qualitative behavior.

We propose that longer periods of reduced gravity would produce a periodic bubble structure in the polymer as can be seen in Figure 6 but which was quickly destroyed by the high g pull out. However, we hasten to point out that large variations in the acceleration level moved the bubbles up and down the tube so that it is impossible to offer a definitive interpretation; longer periods of higher quality weightlessness would be required to test our proposed mechanism.

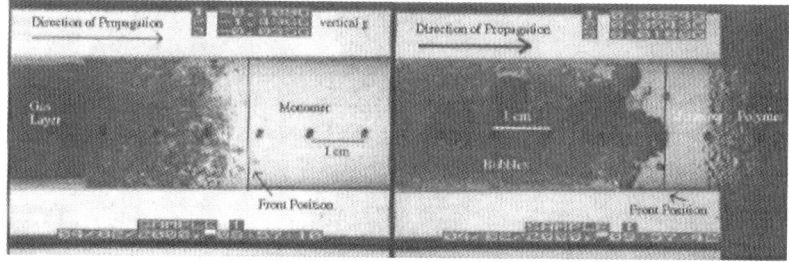

Figure 6. Bubble behavior in liquid/liquid system under (left) high gravity and (right) low gravity.

Viscous Liquid/Liquid System

One way to minimize the effects of g-jitter is to increase the monomer viscosity. We did this by dissolving poly(hexyl acrylate) in the hexyl acrylate. This increased the viscosity of the monomer sufficiently to damp the effects of g-jitter. It also lowered the front temperature. Lowering the front temperature prevents the formation of extraneous bubbles due to dissolved gases and solvents.

As the front nears a region of monomer, the temperature begins to increase rapidly, which rapidly decreases the viscosity. This allows for a smaller region of convection where we can observe bubble migration near the front. When the front reaches that region, the viscosity begins to increase as the monomer is converted to polymer. The bubbles are still able to move in this region but as the front passes, the temperature begins to decrease causing a further increase in the viscosity of the polymer. The viscosity gradient provides a means for some very interesting dynamic properties of both propagating fronts and bubble interaction.

From the KC-135 experiments, we did see the thermocapillary migration of the bubbles to the center of the test tube and since the viscosity of the mixture was much higher, convective migration is limited to the immediate region around the front. Bubble coalescence also slowed. From this we observed that the bubbles formed in reduced gravity would grow until the regions surrounding those bubbles were devoid of gases. As can be seen in Figure 8, the large bubble produced in low gravity remains approximately centered in the tube. When high gravity is again achieved, buoyancy forces the large bubble up through the polymer and into the gas layer.

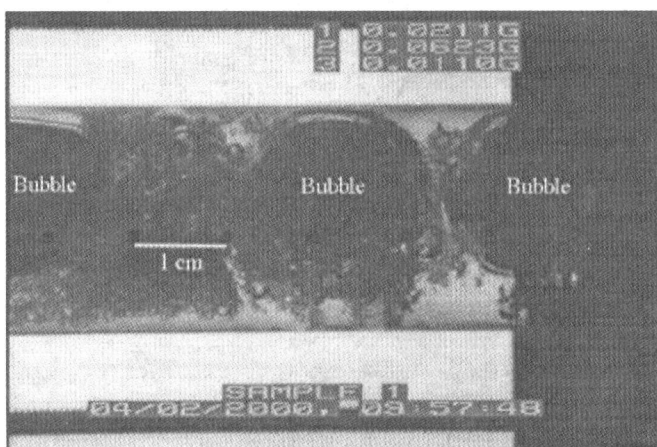

Figure 7. The formation of a transient periodic bubble structure in the liquid/liquid system

Conclusions

Frontal polymerization may provide a viable method for producing foams in space. These KC-135 flights clearly indicate that the acceleration level significantly affects the bubble behavior in a front. The results support the hypothesis that the periodic structure observed on the *Conquest I* sounding rocket was the result of post-front growth and aggregation and not a periodic phenomenon in the front itself. However, additional experiments in longer periods of higher quality microgravity are necessary to definitively test our proposed mechanism.

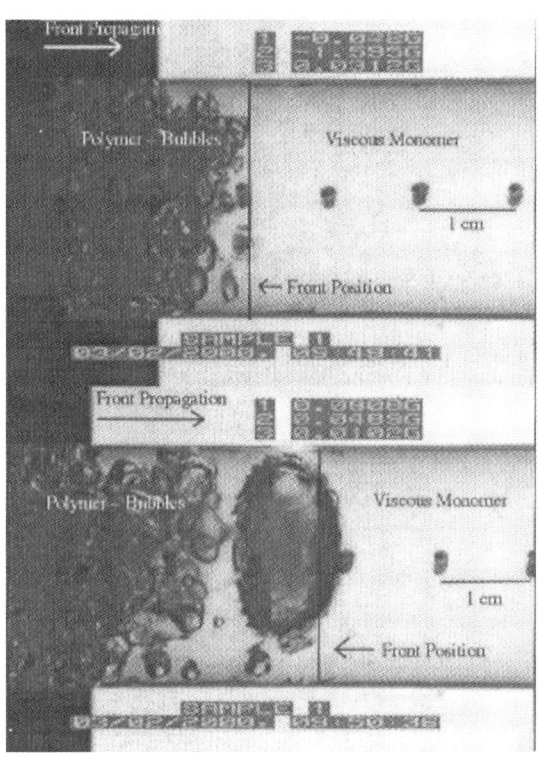

Figure 8. Viscous liquid/liquid system in high (1.6) g (top) and low(0.05) g (bottom)

Acknowledgements

We wish to thank the Department of Chemistry and Biochemistry at the University of Southern Mississippi for its support of this project. We would also like to acknowledge the NASA Reduced Gravity Materials Science Program (NAG 8-973). We would especially like to thank Brent Beabout, Zena Hester, Steve Spearman and Doug Westra for flying the hardware developed at the Marshall Space Flight Center, Steve Fischer, Timothy Dowling and Jeri Briscoe for their efforts in developing the hardware and Jim Bridges and Steve Selph from the USM Fabrication Center for their assistance.

References

1. Bruins, F. P. *Polyurethane Technology*; Bruins, F. P.; Wiley-Interscience: New York, 1969.
2. Saunders, J. H.; Frisch, K. C. *Polyurethanes: Chemistry and Technology, Part 1, Chemistry"*; Wiley-Interscience, New York: 1962.
3. Khemani, K. C. *Polymeric foams: science and technology*; Khemani, K. C.; American Chemical Society: Washington, D.C., 1997.
4. Wessling, F. C.; Maybee, G. W. *Journal of Spacecraft and Rockets* **1989**, *26*, 343-351.
5. Wessling, F. C.; McManus, S. P.; Matthews, J.; Patel, D. *J. Spacecraft* **1990**, *27*, 324-329.
6. Bergman, A.; Frederikson, H.; Shahani, H. *J. Mat. Sci.* **1998**, *23*, 1573-1579.
7. Curtin, M. M.; Tyler, F. S.; Wilkinson, D. *J. Cell. Plastics* **1992**, *28*, 536-556.
8. Chechilo, N. M.; Khvilivitskii, R. J.; Enikolopyan, N. S. *Dokl. Akad. Nauk SSSR* **1972**, *204*, 1180-1181.
9. Davtyan, S. P.; Zhirkov, P. V.; Vol'fson, S. A. *Russ. Chem. Rev.* **1984**, *53*, 150-163.
10. Pojman, J. A. *J. Am. Chem. Soc.* **1991**, *113*, 6284-6286.
11. Pojman, J. A.; Craven, R.; Khan, A.; West, W. *J. Phys. Chem.* **1992**, *96*, 7466-7472.
12. Pojman, J. A.; Willis, J.; Fortenberry, D.; Ilyashenko, V.; Khan, A. *J. Polym. Sci. Part A: Polym Chem.* **1995**, *33*, 643-652.
13. Pojman, J. A.; Nagy, I. P.; Salter, C. *J. Am. Chem. Soc.* **1993**, *115*, 11044-11045.
14. Fortenberry, D. I.; Pojman, J. A. *J. Polym. Sci. Part A: Polym Chem.* **2000**, *38*, 1129-1135.

15. Pojman, J. A.; Curtis, G.; Ilyashenko, V. M. *J. Am. Chem. Soc.* **1996**, *115*, 3783-3784.
16. Pojman, J. A.; Ilyashenko, V. M.; Khan, A. M. *J. Chem. Soc. Faraday Trans.* **1996**, *92*, 2825-2837.
17. Khan, A. M.; Pojman, J. A. *Trends Polym. Sci. (Cambridge, U.K.)* **1996**, *4*, 253-257.
18. Pojman, J. A.; McCardle, T. W. U.S. Patent 6,067,406,2000
19. Bowden, G.; Garbey, M.; Ilyashenko, V. M.; Pojman, J. A.; Solovyov, S.; Taik, A.; Volpert, V. *J. Phys. Chem. B* **1997**, *101*, 678-686.
20. McCaughey, B.; Pojman, J. A.; Simmons, C.; Volpert, V. A. *Chaos* **1998**, *8*, 520-529.
21. Pojman, J. A.; Khan, A. M.; Mathias, L. J. *Microgravity sci. technol.* **1997**, *X*, 36-40.
22. Pojman, J. A.; Khan, A. M.; Mathias, L. *Polym. Prepr. (Am Chem. Soc. Div. Polym. Chem.)* **1997**, *38(1)*, 717-718.
23. Pojman, J. A. Frontal Polymerization in Microgravity, AIAA 98-08113, *36th Aerospace Sciences Meeting,* **1998**,
24. Pojman, J.; Volpert, V.; Dumont, T.; Ainsworth, W.; Chekanov, Y.; Masere, J.; Wilke, H. Bubble Behavior and Convection in Frontal Polymerization on the KC-135 Aircraft, AIAA 2000-0856, *38th Aerospace Sciences,* **2000**,
25. Pojman, J. A.; Ilyashenko, V. M.; Khan, A. M. *Physica D* **1995**, *84*, 260-268.
26. Chekanov, Y.; Pojman, J. A.; Rasmussen, J. K. *Polym. Prepr. (Am Chem. Soc. Div. Polym. Chem.)* **1998**, *39*, 465-467.
27. Balasubramaniam, R.; Lacy, C. E.; Woniak, G.; Subramanian, R. S. *Phys. Fluids* **1996**, *8*, 872-880.
28. Probstein, R. F. *Physicochemical Hydrodynamics.*; Wiley: New York, 1994.

Chapter 9

Instrumentation for Studying Polymer Film Formation in Low Gravity

Paul Todd[1], Matthew R. Pekny[2], Jeremiah Zartman[3], William B. Krantz[4], and Alan R. Greenberg[5]

[1]SHOT, Inc., 7200 Highway 150, Greenville, IN 47124
[2]Intel, Inc., San Jose, CA 95054
[3]Department of Chemical Engineering, University of Colorado, Boulder, CO 80309–0424
[4]Department of Chemical Engineering, University of Cincinnati, Cincinnati, OH 45221–0170
[5]Department of Mechanical Engineering, University of Colorado, Boulder, CO 80309–0427

Tools for investigating polymer film formation in low gravity on spacecraft and aircraft are described. Specific design requirements include rapid phase demixing for short-duration low gravity, vapor phase transport of the evaporating solvent in low gravity, and real-time optical monitoring of the demixing process.

The initial impetus for the commercial membrane industry was seawater desalination, which today is used to provide 2 billion gallons per day worldwide. The energy crisis of the 1970's created new markets for membranes. Since membrane separations do not necessarily require high temperatures or phase change, they offer the possibility of energy-efficient separations. The new technologies of the 1980's created ever-broadening markets for the membrane industry. For example, the semiconductor industry uses membranes to supply the ultrapure water required for rinsing operations involved in the manufacture of wafers and integrated circuits. The rapidly emerging biotechnology industry employs membranes to separate materials sensitive to heat, phase change and other denaturing influences associated with conventional separations; in addition, it employs membranes as carriers for immobilized enzymes for use in biochemical reactors.

This explosion in the membrane market has created intense worldwide competitiveness in the membrane industry. Leadership in this technology requires that membranes be developed rapidly in response to newly defined needs. The morphology or structure of polymeric membranes is pivotal in imparting their unique selective permeation properties. This morphology in turn is created during the membrane-formation process. It is surprising that in most cases our knowledge of membrane morphology is so incomplete that a one-to-one correspondence between the particular steps of a commercial membrane fabrication process and specific membrane characteristics cannot be established.

The majority of polymeric membranes can be formed via a phase separation/inversion process by which a polymer solution (in which the solvent is the continuous phase) inverts into a swollen three-dimensional macromolecular network or gel (where the polymer is the continuous phase). Phase separation can be induced by solvent evaporation (the dry-cast process), nonsolvent/solvent exchange (the wet-cast process), cooling (the thermal-cast process), and polymer leaching (the polymer-assisted process).

The dry-cast process for polymeric membrane formation involves dissolving the polymer in an appropriate volatile solvent containing a small amount of nonsolvent to form a single-phase solution. Subsequent evaporation of the solvent causes a phase separation to occur at a sufficiently low solvent concentration. The resulting nonsolvent-rich dispersed phase forms the pores, whereas the polymer-rich phase forms the matrix structure of the membrane. Thus, in the cellulose acetate/acetone/water system, acetone evaporates, cellulose acetate (CA) becomes the continuous phase, and water forms the pores.

Macrovoid (MV) pores are observed in both the dry-cast and wet-cast processes. MVs are large-scale defects with a characteristic length that can approach the membrane thickness. MVs are generally undesirable because they decrease the permselective performance of membranes in applications such as microfiltration, ultrafiltration and reverse osmosis.

Unique Opportunities Offered by Casting Membranes in Low-g

The growth of macrovoids can be studied using low-gravity methods. Two hypotheses have been offered to explain the growth of MVs (*1*), one of which, the water-diffusion hypothesis, does not involve the significant influence of any body forces; in contrast, the solutocapillary (Marangoni flow) hypothesis involves a mechanism that is strongly influenced by body forces. Hence, casting polymeric membranes in low-g should permit discriminating between these two mechanisms. In particular, by reducing the gravitational acceleration, the resisting buoyancy forces will be reduced; this should result in more rapid MV growth which should be verifiable via greater MV penetration into the casting solution and possibly via an altered MV shape. Moreover, by reducing the buoyancy forces, it should be possible to cause MV growth under casting-solution conditions that do not result in MV growth at 1 g. Observing the latter would provide a particularly discriminating test of these two hypotheses.

MV growth is an example of several questions that can be addressed by membrane processing in low gravity. It is quite possible that entirely new membrane structures may be created in low-g. Indeed, such low-g studies may lead to modified manufacturing practices that can be used to control asymmetric membrane morphologies.

This article addresses the physical problems that are encountered when membrane casting is practiced in low gravity and emphasizes the instrumentation designed to overcome the expected physical challenges of the microgravity environment. The seemingly simplest of these challenges is the complete containment of all fluids. Another is control of the vapor-phase transport of the evaporating solvent. Finally, knowing that phase inversion occurs during low g is important, and optical reflectometry can provide this information.

Prior Studies of Membrane Casting in Low-g

To our knowledge the first membrane-casting experiments in low-g spaceflight were performed by Vera, Pellegrino, Stewart and Todd on the Consort 1 sounding rocket flight (*2,3*), on which CA, polysulfone, chitosan and alumina systems were studied. The early findings, reported informally (*4,5,6*), were limited to the observation that one set of conditions led to the formation of a $1-\mu$ m film of Nafion in low-g using a system that forms a $7-\mu$ m symmetric membrane under laboratory conditions. In some cases the membranes were cast on the surface of an alumina microporous filter, rather than on a smooth glass surface, so that permeability tests could be carried out. I. Vera also cast CA ultrafiltration membranes during the 6-min low-g period on the Consort 4 sounding-rocket flight in 1992.

Vera (*2*) designed experiments to be performed in a "Get Away Special" (GAS) canister on the U. S. space shuttle. The apparatus consisted of a pair of hemicylinders in one of which a large (15 x 40 cm) cellulose acetate membrane was to be wet-cast on a polished surface and quenched with water pumped from the opposite hemicylinder. This experiment would have resulted in the largest polymer membrane cast in space to date; however, the experiment, scheduled for flight at the time of the Challenger tragedy, was not performed in space.

Automatic Membrane Casting Devices

The rocket-borne experiments were performed using an automated minilab, the Materials Dispersion Apparatus (MDA) (Fig. 1), which consists of two blocks of inert plastic (white Delrin) with multiple wells drilled at precise intervals in them. It functions by bringing wells in opposite blocks into contact with each other in response to an electronic signal. The version used in the sounding-rocket experiments also accommodated several other fluids and bioprocessing experiments (3,7). Three wells in an upper block contained, sequentially, polymer solution, air and water. The air well was vented to allow mass transfer of evaporating solvent. Before launch a very shallow lower well with a glass or ceramic surface was in contact with the polymer solution, and the thin film, with a diameter of 0.64 cm, would form in this well after activation. Activation occurred in response to a low-gravity signal, and the shallow well was moved to the vented air cavity so that solvent evaporation could be initiated. After a predetermined number of minutes the block moved once more, and water was contacted with the nascent membrane leading to the standard quenching process used in membrane wet-casting. It is estimated that Reynolds' number is less than 10 for the fluid motion as wells come into contact. As solvent is transported out of the phase-changing film into the water, convective processes are thought to play a role (8). If so, the nature of the resulting membranes should be modified in low gravity where density gradients do not lead to natural convection.

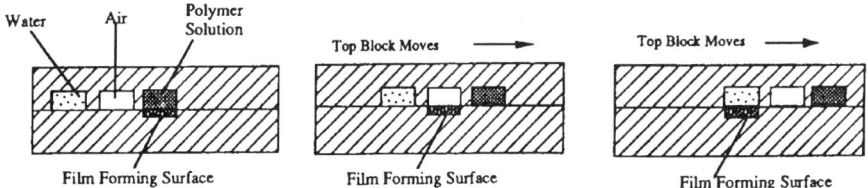

Figure 1. Initial sliding-block mechanism for casting polymer and ceramic membranes in low-gravity (9).

Dry-casting experiments were also performed in this device. In this case, the water step was omitted. Gamma-alumina ceramic membranes were cast by the sol-gel process, and chitosan and composite membranes were cast by the dry-cast phase-inversion process. In the case of the chitosan membrane, a dense film was observed where a microporous membrane would have been expected in 1 g (4). Attempts were made to cast five different membrane compositions on a sounding rocket flight (3).

A membrane-casting apparatus was designed and constructed at the University of Colorado, originally for flight on the ill-fated Joust 1 mission. The important feature of this device was its arrangement for transport of the evaporating acetone. This was accomplished by placing a layer of activated carbon above the casting surface. The rapid absorption of the acetone created a strong driving force (normally provided by buoyant force on the ground) to maintain the flow of acetone vapor away from the film surface. This approach had been already tested on the ground. Independently and for a different experiment, Lichtenbelt et al. (10) successfully demonstrated this method of acetone vapor transport in low-g. This method of acetone vapor transport therefore was to be used in the apparatus design, the principle of which is shown in Figure 2. The absorption of acetone vapor in the activated carbon also prevents the escape of evaporated acetone, a potential safety hazard, into the flight vehicle environment.

Battelle Memorial Institute, using the apparatus of Cassanto et al. (3) on KC-135 flights, established conditions for casting polymer membranes on later orbital space flights. Experiments were conducted by Battelle Memorial Institute on the Space Shuttle through sponsorship of the former NASA Office of Commercial Programs. These consisted of a simple two-chamber vacuum casting device that first layered polymer solution on the substrate, evacuated the chamber to draw away the evaporation products, then repressurized the chamber to one atmosphere (11).

A sliding-block casting apparatus was designed and constructed for fabricating polymeric membranes in the KC-135 aircraft, based on the sounding-rocket design as shown in Figure 2. This apparatus was fitted with an optical phase-change sensor which permits obtaining real-time data during low-g for the phase-change process that occurs during membrane formation. Ground-based experiments were done using this apparatus and optical sensor to serve as the control for comparison with the low-g studies. Sub-orbital experiments occurred on the KC-135 aircraft in 1995, 1997 and 1999. A chimney full of polymer solution is held over a casting well 100-250 µm deep. In the MCA-1 the polymer solution upon triggering is translated by motion of the lower sliding block to a position beneath a chimney full of activated carbon. At this point solvent evaporation begins and will continue until evaporation is complete. We designed and constructed a modified version of the MCA-1 for studying MV growth in the dry-casting process. This modified version of the MCA-1 has the capacity to cast six membranes in a single motion of a sliding block and has the capability for real-time observation of the inception and duration of the phase-separation process that leads to membrane formation (*12*). The principle of this system is diagrammed in Figure 2. One of the six membrane-casting cells is interfaced to the real-time optical monitoring system.

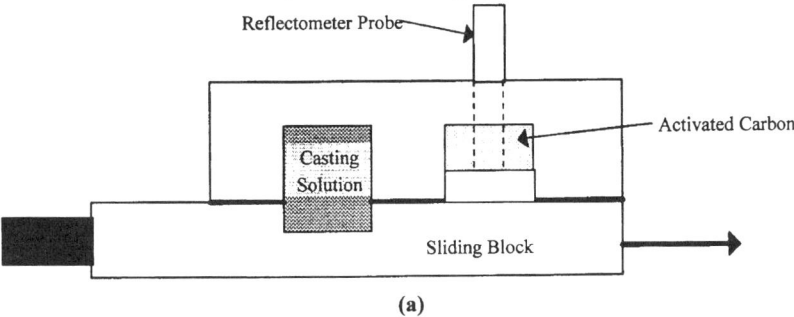

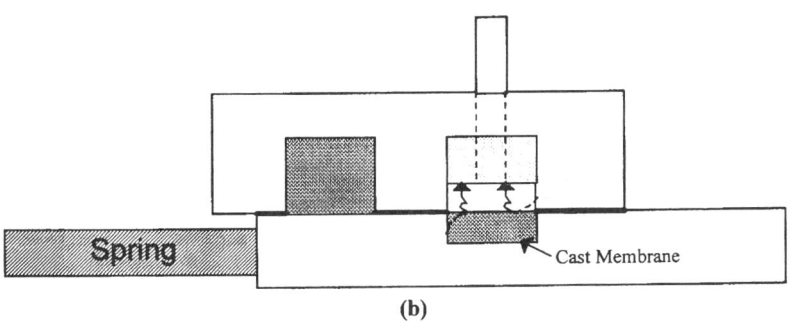

Figure 2. Sketch showing assembled construction of a casting station of the Membrane Casting Apparatus (MCA-1) with (a) opposing chambers positioned for loading and (b) positioned for membrane casting. The left chimney is filled with casting solution, and the right chimney contains activated charcoal adsorbent.

The principle of Figure 2 has been implemented in the overall design of the MCA-1 as shown in Figure 3, which indicates how six casting stations are configured and how a fiber optic cable is attached to one of them. The Cocking Mechanism slides the Sliding Block to the right, compressing the Spring-Loaded Plungers. Upon removing the Firing Pin (upward) the Sliding Block is released and slides rapidly to the left, sweeping along the shallow well (Fig. 2) from the solution-containing chimney to the carbon-containing chimney. Each MCA-1 casts 6 membranes with each activation, and the progress of one membrane through the demixing stages is monitored optically through the Fiber Optic Holder. The Chimney Block and the Base Plate do not move.

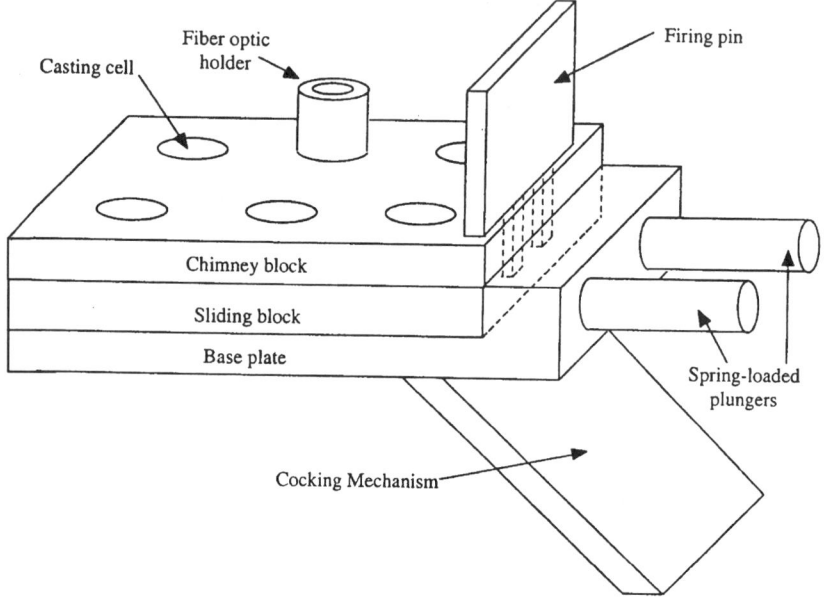

Figure 3. Schematic diagram of a Membrane Casting Apparatus (MCA-1) for application in microgravity.

It can be discerned from Figure 3 that the sliding block containing the membrane solution will move very rapidly to its stopping point (in a few milliseconds). This time is brief compared to the low-g time available, which is an advantage. However, by design the membrane solution is still liquid at the end of the excursion and tends to "slosh" (Figure 4). This action results in non-uniform membrane thickness after casting and biases the occurrence of macrovoids. If "sloshing" is allowed, macrovoids will not be distributed homogeneously in the resulting membrane, since the membrane thickness is not uniform. This situation leads to data sampling problems when membrane fragments are examined by electron microscopy. One solution to the "sloshing" problem is elimination of the spring-loaded plungers. It has been demonstrated that moving the sliding block by hand over a period not exceeding one second can result in improved distribution of macrovoids (*13*). This is accomplished by attaching a handle to the sliding block identified in Figure 3. Very little practice is required to operate this version of the MCA properly and, during flight, the operator usually releases the entire assembly shown in Figure 3 so that it is "floating" inside the aircraft. It is estimated that the maximum acceleration during phase separation under these conditions is around 0.001 g.

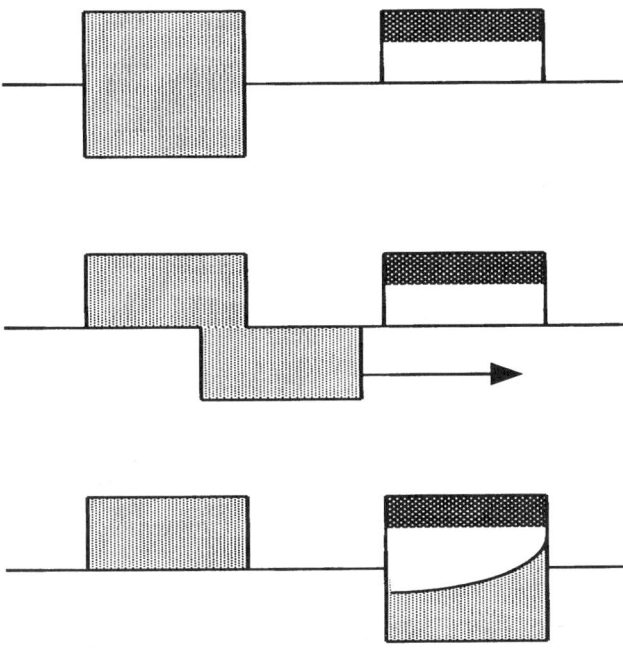

Figure 4. "Sloshing" of polymer solution before phase inversion occurs owing to excessive deceleration after the Sliding Block has been moved rapidly by the Spring Plungers.

The design for a self-contained optical sensor-recorder for the MCA-1 has been accomplished and implemented. Figure 5 indicates the external design consisting of a power supply, photodiode, filament lamp (or light-emitting diode or diode laser), high-impedance operational amplifier, and a clock and memory circuit programmed into a personal computer. This system is designed to record reflected light intensity for the duration of the casting process and to store intensity-vs-time points in digital form. The intensity of light reflected from the thin film changes during phase demixing and stops changing when phase inversion has ended. The optical intensity data are initially written to RAM to avoid reliance on disk-drive operations during low g; the data are subsequently transferred to the computer's hard disk. The integrated unit has been prepared for and meets requirements of KC-135 low-g flights (*14*). This system has been used for the casting of about 200 membrane samples in such low-g flights. An example of an optical data set, showing the onset and completion of phase separation is given in Figure 6.

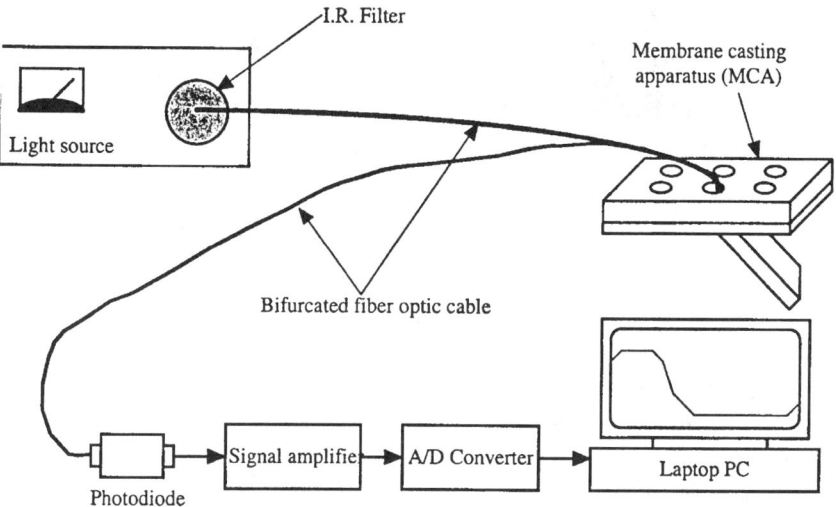

Figure 5: Schematic sketch of an optical reflectometry system. The activated-carbon chimney is perforated to admit the optical fiber (or laser beam directly when used in interferometry mode). The detector used in reflectometry is a photodiode coupled to a high-impedance pre-amplifier, the signal from which is passed through the A/D converter to the digital recorder.

Summary and Conclusions

Methodology and instrumentation for creating small samples of polymer membranes cast in low gravity have been produced and operated in low gravity. Physical problems that have been addressed include the complete containment of all fluids including the evaporating solvent vapor, mass transfer of solvent vapor away from the membrane surface, the monitoring of phase changes by reflectometry and the selection of appropriate illumination for reflectometry during shifts in inertial acceleration. These problems were solved, respectively, by using a fully enclosed sliding block device that can be operated manually or automatically, absorbing the solvent vapor into activated carbon at a mass transfer rate similar to that caused by convective removal into an infinite volume at 1 g, computerized recording of light intensity vs. time to assure the monitoring of demixing in short times (< 20 sec), and using a diode laser in place of a filament lamp as illuminator of the polymer solution.

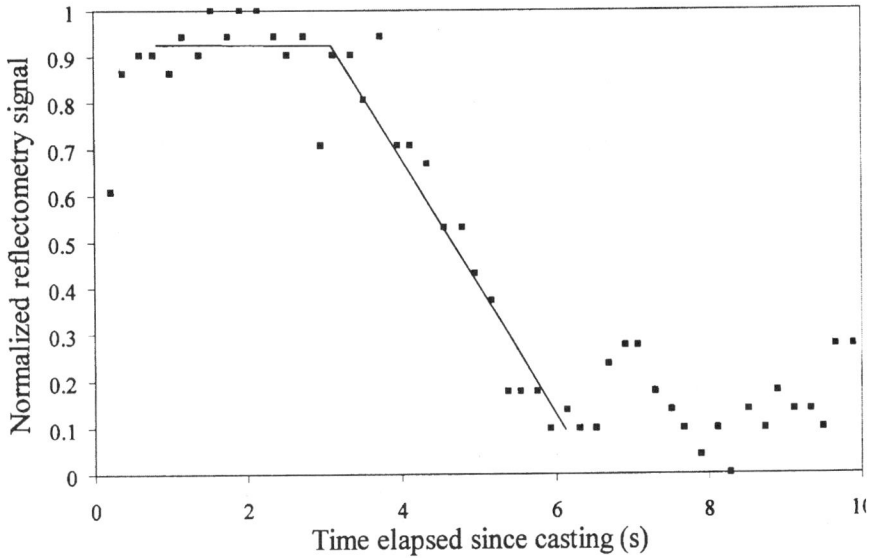

Figure 6. Light reflectance vs. time recorded using the system shown in Figure 5 with a filament-lamp illuminator at 1 g. The casting-solution composition was 10 % cellulose acetate, 27 % water, 63 % acetone (wt/wt). The initial thickness of the solution was 150 μm. The solid line indicates the drop in reflectance due to demixing (phase separation). Demixing begins after about 3.5 seconds and lasts approximately 3 seconds.

Future needs for instrumentation in the field are many. Longer low-g periods will be needed in order to assess a fuller breadth of problems. Nearly all of the low-gravity observations to date have been limited to the cellulose acetate-acetone-water system. Other polymer solutions will require more or less time to demix. A majority of work has focused on dry casting, although early apparatus was designed around wet casting. Extended MCA designs will be required in order to:

- Function automatically on orbital spacecraft
- Collect larger membrane areas that can be subjected to performance tests
- Perform real-time flow visualization at the microscopic level
- Automatically process large numbers of samples and preserve them

Finally, the siginificance of discoveries that can be made by studying membrane formation in low gravity merits the development of flight equipment capable of fulfilling these needs.

Acknowledgments

This work was supported by U. S. National Aeronautics and Space Administration grants NAG8-1062 and NAG8-1475. We are grateful for the excellent machining services of Mr. Willy Grothe and the flight engineering services of Mr. Mark Rupert of BioServe Space Technologies Center for Space Commercialization.

References

1. Konagurthu, S.; Krantz, W. B.; Todd, P. *Proc. of Euromembrane '95*, Vol. 1, University of Bath, Bath, England, 1996, pp. 256-261.
2. Vera, I. *Adv. Space Res.* **1986,** *6* (5) 65-68.
3. Cassanto,J. M.; Holemans, W.; Moller, T.; Todd, P.; Stewart, R. M.; Korszun, Z. R. *Progr. Astronaut. Aeronaut. 127*: *Space Commercialization: Platforms and Processing*, Shahrokhi, F.; Hazelrigg, G.; Bayuzick, R., Eds.; Amer. Inst. Aeronaut. Astronaut., Washington, DC, 1990, pp. 199-213.
4. Wessling, F.; Lundquist, C.; Maybee, G. *Acta Astronautica*, **1990,** *21*(9), 647-657.
5. Todd, P. *Sounding Rocket Experiments in Biotechnology Using Materials Dispersion Apparatus.* Final Report, Contract No. H8057B, Marshall Space Flight Center, Huntsville, AL, 1989.
6. Pellegrino, J. J.; Todd, P. *International Conference on Physicochemical Hydrodynamics*, MIT, Cambridge, MA, June 25-29, 1989.
7. Todd, P. In *Progress in Low-Gravity Fluid Dynamics and Transport Phenomena.* Koster, J. N.; Sani, R. L., Eds.; *Progress in Astronautics and Aeronautics* vol. 130, American Institute of Aeronautics and Astronautics, Washington, DC, 1990; pp. 539-602.
8. Cabasso, I. ACS Symp. Ser. 153. *Synthetic Membranes: Desalination* . Tubark, A. R., Ed.;American Chemical Society, Washington, DC, 1981; pp. 267-291.
9. Holemans, J.; Cassanto, J. M.; Moller, T. W.; Cassanto, V. A.; Rose, A.; Luttges, M.; Morrison, D.; P. Todd, P.; R. Stewart, R.; R. Z. Korszun, Z. R.; Deardorff, G. *Micrograv. Q.* **1991,** *1*, 235-247.

10. Lichtenbelt, J. H.; Drinkenberg, A. A. H.; Dijkstra, H. A. in *Scientific Results of the German Spacelab Mission D1*, European Space Agency, Paris,1986; p.127.
11. McCauley, L., *in Space Shuttle Mission STS-43 Press Kit*, NASA Kennedy Space Center, July 19, 1991.
12. Konagurthu, S. Ph.D. Dissertation, University of Colorado, Boulder, CO, 1998.
13. Pekny, M. M. S. Thesis, University of Colorado, Boulder, CO, 1999.
14. Pekny, M. R.; Zartman, J.; Greenberg, A. R.; Krantz, W. B.; Todd, P. *Polymer Preprints* **2000**, *41*(1), 1060-1061. American Chemical Society, Washington, DC.

Chapter 10

Liquid Crystal and Polymer Dispersions in a Microgravity Environment

Joe B. Whitehead, Jr.[1], and Gregory P. Crawford[2]

[1]Departments of Physics & Astronomy and Chemistry & Biochemistry, University of Southern Mississippi, 2904 West Fourth Street, Hattiesburg, MS 39406
[2]Division of Engineering, Brown University, 182 Hope Street, Providence, RI 02912

Liquid crystal and polymer dispersions (LCPD) fabricated using polymerization induced phase separation (PIPS) contain morphologies that range from a dispersion of micrometer-sized droplets of a liquid crystal in a solid polymer matrix to a polymer network dispersed in the liquid crystal. The interest in LCPDs continues to grow with applications that range from light shutters to flat panel displays, and to high-resolution projection displays. Herein, we investigate the fabrication of LCPDs processed in terrestrial and microgravity environments. The materials used in this investigation include the liquid crystal mixture E7, the single component liquid crystal K15; the thiol-ene based pre-polymer NOA65, and the liquid crystal monomer RM257. Scanning electron microscopy and laser-light transmission measurements were used to probe the sample morphology and electro-optic parameters of LCPDs processed in the terrestrial and microgravity environments. The results discussed here stem from four parabolic flights spanning the years of 1995-1997.

Introduction

Liquid crystal and polymer dispersions (LCPDs) are promising new materials for anti-reflective coatings, omni-directional reflectors, electrically switchable Bragg gratings, spatial light modulators, optical interconnects, optical data storage, dynamically variable lenses, and high intensity laser radiation attenuators. LCPDs are materials that result from the phase separation of liquid crystalline and polymeric materials that when sandwiched between transparent conducting electrodes maybe electrically switched between scattering and non-scattering states (*1,2*). LCPDs have been developed for a host of electro-optic applications which include low power reflective displays (*3*) dichroic displays (*4*), privacy windows (*5*), solar control windows (*5*), projection devices (*6*), laser projection (*7*), scattering polarizers (*5*), IR shutters (*8*), spatial light modulators (*9*), bi-stable memory devices (*10*), and passive optical elements (*10*). An in-depth understanding of the relationship between the material properties, processing conditions, and the evolution of the phase separated morphology is required to design and fabricate LCPDs to meet the requirements for the aforementioned potential applications. In addition, LCPDs provide a unique opportunity to investigate, on a fundamental level, the thermodynamics and kinetics of the coupled phase separation and polymerization associated with the fabrication of LCPDs. The focus of this chapter is polymer dispersed liquid crystals (PDLC) and network stabilized liquid crystals (NSLC). PDLCs contain a dispersion of micrometer-sized liquid crystal droplets in a solid polymer matrix, and NSLC contain a polymer network dispersed in the continuous liquid crystal.

The morphology of LCPDs ranges from liquid crystal droplets dispersed in a solid polymer for polymer dispersed liquid crystals (PDLC) to a self-assembled polymer network dispersed in the liquid crystal for network stabilized liquid crystals (NSLC). PDLCs contain micrometer-sized liquid crystal droplets randomly dispersed in a solid polymer matrix (see Figure 1). More recently, PDLCs fabricated with holographic processing (H-PDLC) yield alternating planes of nanometer-sized liquid crystal droplets and solid polymer planes. The phase-separated morphology of a holographically formed PDLC (H-PDLC) consists of alternating planes of nanometer-sized liquid crystal droplets and solid polymer (see Figure 2).

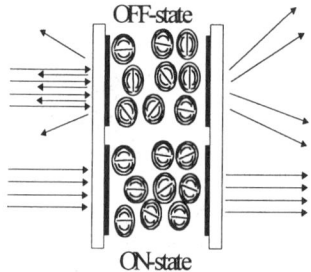

Figure 1. Schematic representation of a polymer dispersed liquid crystal (PDLC) sandwiched between transparent conducting electrodes. The upper pixel is the opaque OFF-state and the lower pixel is the transparent ON-state.

The phase-separated morphology of a NSLC consists of a self-assembled polymer network dispersed in the continuous liquid crystal (See Figure 3). NSLCs differ from both PDLC and H-PDLC materials, in that, the initial mixture contains less than 10% by weight of a reactive liquid crystalline monomer; whereas, the initial mixture for PDLCs and H-PDLCs contains 20% to 60% reactive non-liquid crystalline monomer. NSLC are bi-stable in that once switched with an applied voltage from the reflective state to the transparent state or visa versa, the optical properties do not change when the voltage upon removed.

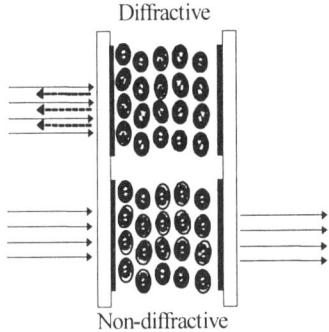

Figure 2. A schematic representation of an H-PDLC sandwiched between transparent conducting electrodes. The upper pixel is diffractive OFF-state and the lower pixel is the non-diffractive ON-state.

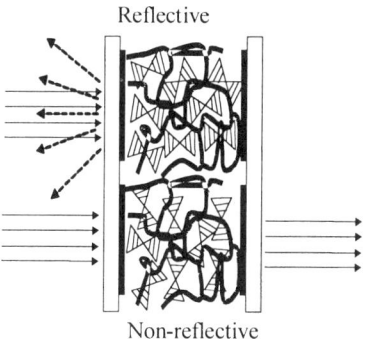

Figure 3. A schematic representation of a NSLC sandwiched between transparent conducting electrodes. The upper pixel is the reflective OFF-state and the lower pixel is the non-reflective ON-state.

Liquid crystal and polymer dispersions are fabricated using thermally-induced phase separation (TIPS), solvent-induced phase separation (SIPS), or Polymerization-induced phase separation (PIPS)(*11*). For TIPS, a homogeneous mixture of a low-molecular weight liquid crystal and thermoplastic polymer is cooled below the critical phase separation temperature to induce phase separation into liquid crystal rich and polymer rich domains. The morphological properties (domain size, number of domains per unit volume, and the composition of the domains) depend primarily on the choice of liquid crystal and thermoplastic polymer, the initial weight fraction of liquid crystal in the initial mixture, and the rate of cooling.

For SIPS, the common solvent for the liquid crystal and thermoplastic or pre-polymer is extracted from the initial mixture to induce phase separation. The morphology depends primarily on the choice of liquid crystal and thermoplastic or pre-polymer, initial weight fraction of liquid crystal in the initial mixture, and the rate of solvent extraction. A major disadvantage of LCPDs produced using TIPS and SIPS is the lack of a thermally stable morphology because the polymer matrix is not crosslinked.

For PIPS, polymerization of the monomer or pre-polymer component in the initial mixture of liquid crystal, monomer, or pre-polymer, and photo-initiator induces phase separation. The conversion of monomer to polymer with thermal or photon initiated polymerization destabilizes the homogeneous mixture and induces phase separation. Photon-initiation affords the flexibility of multi-wavelength processing that has resulted in H-PDLCs and other spatially tailored LCPD morphologies (*12*). The morphology depends primarily on the choice of liquid crystal and monomer/pre-polymer, the weight fraction of liquid crystal in the initial mixture, the cure temperature, and the rate of polymerization. Another crucial factor that influences the phase-separated morphology is whether phase separation occurs before or after gelation of the polymer matrix (13).

Electro-optic Properties of PDLCs

The electro-optic properties (contrast, switching time, switching voltage, etc.) of PDLC materials are determined by the morphology. It is convenient to consider the spatial and compositional aspects of the morphology separately when discussing the electro-optic properties of PDLCs. The spatial morphology refers to the geometrical aspects of the morphology (droplet size, droplet shape, and number of droplets per unit volume); whereas; the compositional morphology refers to the distribution of liquid crystal and polymer molecules within the phase-separated morphology.

For typical values of liquid crystal and polymer refractive indices, PDLCs scatter in the Rayleigh-Gans or Anomalous Diffraction regime (14). In both regimes, the scattering depends on droplet size and droplet size distribution, droplet shape, number of droplets per unit volume (i.e. spatial morphology) and the refractive index difference between the liquid crystal droplets and the polymer matrix (i.e. compositional morphology). A normal mode PDLC is opaque in the OFF-state (upper pixel in Figure 1) and transparent in the ON-state (lower pixel in Figure 1). Therefore, contrast, the ratio of the ON-state and OFF-state transmission, depends on the intrinsic scattering properties of the PDLC. The opaque (scattering) OFF-state is due to the refractive index difference between the liquid crystal droplets and the surrounding polymer matrix. The transparent ON-state results from electrically aligning the liquid crystal droplets with the application of an electric field such that the refractive index of the droplets matches that of the surrounding polymer matrix (lower pixel in Figure 1).

The switching time, the sum of the turn-on and turn-off times, and switching voltage depend on both the compositional and spatial morphologies. The magnitude of the applied voltage and the viscosity of the liquid crystal are the limiting factors for the turn-on time. It is more difficult to distort (i.e. align) small liquid crystal droplets; therefore, the switching voltage is typically inversely proportional to droplet size. In contrast to the switching voltage, the turn-off time scales with droplet size. The switching voltage and turn-off time also depends on the shape of the liquid crystal droplets (15). In addition, the switching voltage depends on the electrical properties of the liquid crystal droplets and surrounding polymer. The electrical properties of the liquid crystal and surrounding polymer matrix are controlled by the compositional morphology. For example, higher applied voltages are required to switch PDLC materials if the resistive losses reduce the electric field experienced by the droplets.

The spatial and compositional morphologies determine the electro-optic properties of PDLC materials. Therefore, a thorough understanding of morphological development is critical precise control of the morphology (spatial and compositional) and therefore, optimization of PDLC electro-optic properties.

Microgravity

Further development of PDLCs requires detailed knowledge of the morphological evolution from the initial homogeneous mixture to the final phase-separated morphology. Microgravity studies provide one avenue to investigate the morphological evolution and to ascertain the influence, if any, of buoyancy-driven convection, and surface tension on the morphology. Microgravity studies, to date, range from experimental drop tower studies and parabolic flights to numerical simulations. In a microgravity environment, buoyancy driven convection has limited influence on PDLC morphology due to the reduction in density difference among the components of the PDLC mixture prior to and during polymerization. Fox and co-workers fabricated PDLCs in a microgravity environment aboard NASA's KC135 aircraft (*16*). The electro-optic performance of the microgravity samples was superior to the corresponding terrestrial samples. The superior electro-optic performance of the microgravity over the terrestrial samples was attributed to the more uniform spatial morphology (i.e. droplets and droplet size distribution) of the microgravity samples.

In addition to the parabolic flights, drop tower experiments and computer simulations have been used to investigate the gravitational influence on PDLC morphology. Jin and co-workers found reasonable agreement between droplet size and droplet size distribution for PDLCs produced via the TIPS process in drop tower experiments and Monte Carlo simulation (*17*). The 2-dimensional character of the simulations limited the comparison between experiment and simulation. In a subsequent investigation, Jin and co-workers again employed Monte Carlo simulations to investigate the effect of polymerization reactivity, interfacial strength, and gravity on PDLCs produce via the PIPS process (*18*). In a low-interfacial strength regime, the collision probability for the liquid crystal droplets is sufficiently high that growth by coalescence occurs in the terrestrial and microgravity environments. Thus, the simulations predict that PDLC morphology is gravitationally depend only in a high-interfacial strength regime where the collision probability varies with gravitational attraction. Parbhakar and co-workers developed a kinetic model for phase separation that predicted a Gaussian shaped droplet size distribution in the absence of coalescence as a growth factor (*19*). Results from drop tower experiments reported in this reference support the results of the kinetic model. In addition, Parbhakar and Dao reported that PDLCs processed in microgravity exhibited higher ON-state transmission, lower threshold voltages, and lower ON-state voltages than the corresponding terrestrial samples (*20*).

Experimental

Liquid crystal and monomer/pre-polymer mixtures containing a low-molecular weight non-reactive liquid crystal and monomer/pre-polymer were prepared to achieve the PDLC and NSLC morphologies. The mixtures for PDLC morphology contained the multi-component liquid crystal E7 obtained from EM Industries (see Figure 4) and the pre-polymer NOA65 obtained from Norland Optical Adhesives. The mixtures for the NSLC morphology contained the low-molecular weight non-reactive liquid crystal K15 (see Figure 3) and the nematic liquid crystalline monomer RM257 obtained from EM Industries (see

K15: R=$CH_3(CH_2)_4$
K21: R=$CH_3(CH_2)_6$
M24: R=$CH_3(CH_2)_7O$

T15: R=$CH_3(CH_2)_4$

E7 Composition

K15	51%
K21	25%
M24	16%
T15	8%

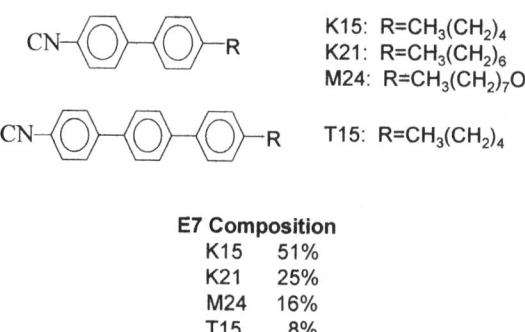

Figure 4. The structures and composition of the nematic liquid crystal mixture known as E7.

Figure 5). All mixtures were prepared by 1) weighing the appropriate quantity of each component, 2) vigorously stirring, and 3) centrifuging (to eliminate air bubbles introduced during step 2). Steps 2 and 3 are repeated to ensure the mixture is homogeneous. The compositions of the E7/NOA65 PDLC mixtures include 35%, 45%, 55%, and 60% by weight. The compositions of the

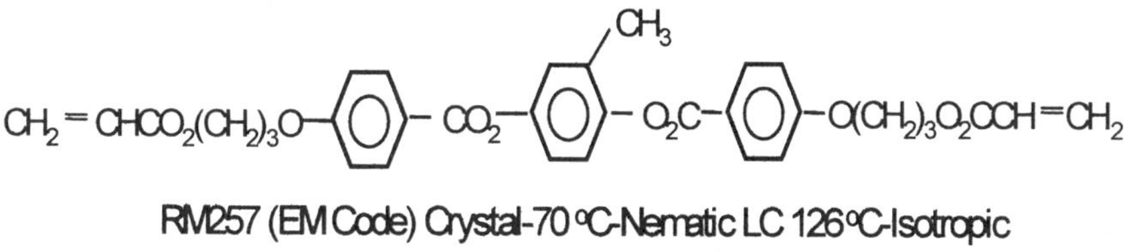

$$CH_2=CHCO_2(CH_2)_3O-\underset{}{\bigcirc}-CO_2-\underset{CH_3}{\bigcirc}-O_2C-\underset{}{\bigcirc}-O(CH_2)_3O_2CCH=CH_2$$

RM257 (EM Code) Crystal-70 °C-Nematic LC 126 °C-Isotropic

Figure 5. Chemical structure for the nematic monomer RM257 from EM Industries.

K15/RM257 NSLC mixtures include 2%, 4%, and 6% by weight. In the case of the PDLC samples, the mixtures were capillary filled in 10 micrometer thick cells constructed from indium-tin-oxide coated glass substrates. For the NSLC samples, the K15/RM257 mixtures were capillary filled into 10 micrometer thick cells constructed from indium-tin-oxide coated glass substrates coated with a uniformly rubbed polyimide layer. The uniformly rubbed polyimide layer macroscopically orients the K15 molecules which in-turn induces the alignment

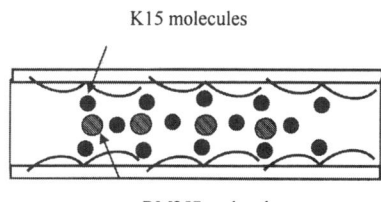

Figure 6. Schematic representation (edge view) of the macroscopic alignment of the RM257 molecules by the K15 molecules due to the rubbed polyimide.

of the RM257 molecules (see Figure 6).

The PDLC and NSLC mixtures were photo-polymerized in the terrestrial and microgravity environments to probe the gravitational influence on the phase-separated morphologies. All samples were photo-polymerized in a reactor designed and built at Marshall Space Flight Center. The reactor consisted of a fluorescent bulb as the ultra-violet source, surrounded by three sample rails with five sample positions on each rail. A rotating shutter was used to shield the samples from the UV radiation before the predetermined exposure period in the microgravity environment. Laboratory characterization of the PDLC and NSLC mixtures using differential photo calorimetery demonstrated that the polymerization reaction was complete within 35 seconds. For samples processed in microgravity, polymerization was initiated 5 seconds before entering the 25-second microgravity portion of the parabolic maneuver. Terrestrial control samples were photo-polymerized in the laboratory in the reactor using the same protocol as for the microgravity samples.

Characterization of the microgravity and terrestrial samples included electro-optic measurements, optical microscopy, and scanning electron microscopy. Electro-optic characterization of the PDLCs samples employed voltage-dependent light transmission. The following equation describes laser light transmission through a PDLC:

$$I_t = I_o e^{-\beta d \sigma}$$

where I_t and I_o are the transmitted and incident intensity, respectively. d is the sample thickness, β is the number of droplets per unit volume, and σ is the voltage dependent droplet scattering cross.

Each sample was subjected to a ramped voltage waveform (1kHz modulation) to identify the voltage, V_{90}, required to achieve 90% of the sample's maximum laser light transmission. The laser light transmission was then measured upon the application of a voltage pulse (1kHz modulation) of constant amplitude corresponding to V_{90}. The switching voltage, contrast, and switching time were extracted from the sample's response to voltage pulse waveform. The electro-optic measurement apparatus is schematically shown in Figure 7.

The NSLC samples were characterized using optical microscopy to probe the macroscopic alignment of the polymer network and scanning electron microscopy to probe the surface of the polymer network.

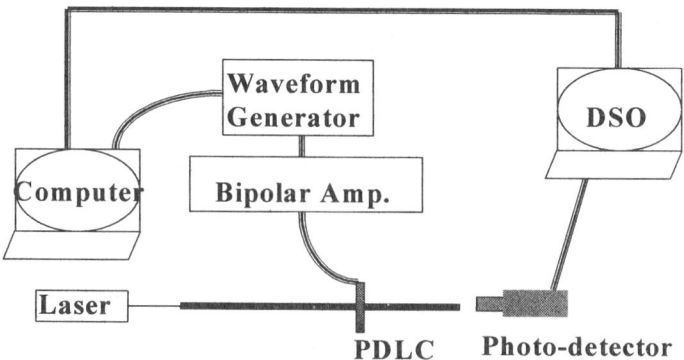

Figure 7. Schematic diagram of the electro-optic measurement apparatus.

RESULTS AND DISCUSSION

The results of the first flight in 1995 were consistent with the results of Fox and co-workers (*16*). The microgravity PDLC samples contained smaller more spherical droplets than the corresponding terrestrial samples. For example, scanning electron analysis of 40% E7/NOA65 samples revealed that the microgravity samples contained droplets less than one micron in diameter, whereas the terrestrial samples contained droplets approximately 2.0 microns in diameter (*21*). The morphological influence on the electro-optic properties is shown in Figures 8 and 9. Figure 8 contains the electro-optic response of a 60% E7/NOA65 PDLC fabricated on earth and Figure 9 contains the electro-optic response for the same PDLC mixture fabricated in microgravity. The terrestrial PDLC had a longer switching time primarily because of the multi-stage decay that includes contrast inversion (the transmitted intensity dipped below the level of the Off-state transmission). This type of behavior is indicative of irregularly

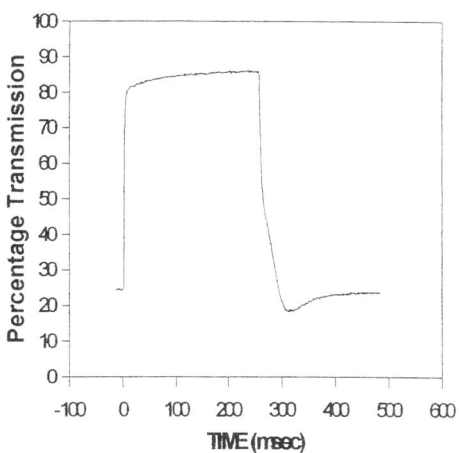

Figure 8. Electro-optic response for a 60% E7/NOA65 sample fabricated in the terrestrial environment (Reproduced with permission from reference 21. Copyright 1997 SPIE).

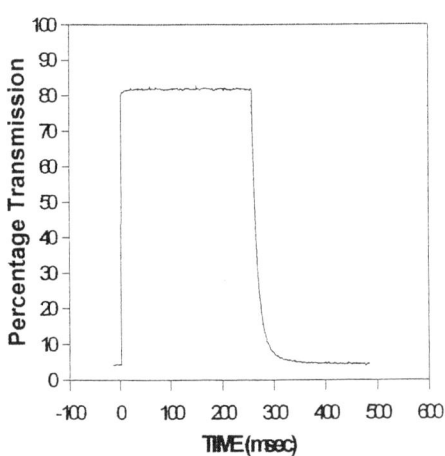

Figure 9. Electro-optic response for a 60% E7/NOA65 sample fabricated in the microgravity environment (Reproduced with permission from reference 21. Copyright 1997 SPIE).

shaped droplets with some degree of connectivity. The contrast for the microgravity PDLC (20:1) is four times larger than contrast (5:1) for the terrestrial PDLC. The difference in contrast is due the lower off-state transmission for the microgravity PDLC that results from increased uniformity in the droplet morphology. The switching voltage for the terrestrial and microgravity PDLCs were 1.1 volts/micron and 1.8 volts/micron, respectively. It should noted that 60% E7 in NOA65 is close the solubility limit for the unpolymerized mixture at room temperature.

On a subsequent flight, 45% E7/NOA65 PDLCs were fabricated in microgravity. The electro-optic responses for the terrestrial and microgravity PDLCs are shown in Figure 10. The switching time for the terrestrial PDLC was 29 milliseconds; whereas, the switching time for the microgravity PDLC was 14 milliseconds. The switching voltage, V_{90}, was 5.7 volts/micron and 5.3 volts/micron for the terrestrial and microgravity PDLCs, respectively. Contrast for the terrestrial and microgravity samples were 8.5 and 2.7, respectively. For these 45% E7/NOA65 PDLCs the switching time was improved with microgravity processing, but the contrast was not. The contrast, switching time and switching voltage results are consistent with the terrestrial PDLC containing a broader droplet size distribution, less spherical droplets, and a higher number of droplets per unit volume than the corresponding microgravity PDLC.

The NSLC samples processed in microgravity on the 1997 KC-135 flight were characterized using optical microscopy and scanning electron microscopy. Unlike, the PDLC results that stemmed from multiple microgravity flights, the NSLC observations described below are preliminary and qualitative. After

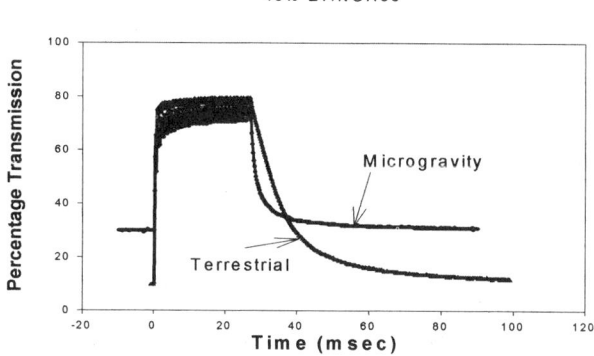

Figure 10. Voltage-dependent laser-light transmission for 45% E7/NOA65 samples fabricated in microgravity and terrestrial environments.

polymerization in the terrestrial or microgravity environment, each NSLC sample was heated above the nematic-to-isotropic temperature to observe t the presence of the polymer network. Polymer networks were not observed in NSLC samples with less than 6% RM257 in the liquid crystal K15 regardless of the gravitational environment. For the NSLC samples with polymer networks, the microgravity samples appear to retain more of the pre-polymerization macroscopic alignment than the corresponding terrestrial samples (see Figure 11). Scanning electron microscopy revealed differences in the surface texture of the terrestrial and microgravity polymer networks. The high-resolution scanning electron images are shown in Figure 12.

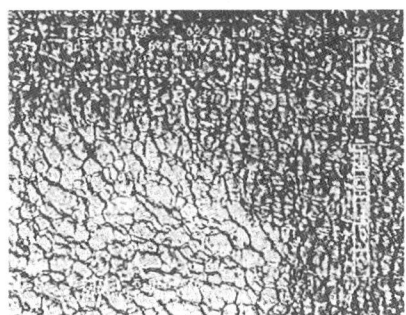
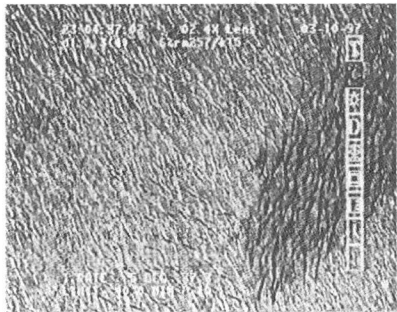

Figure 11. Optical micrographs of the polymer network contained in NSLC (6% E7/RM257) fabricated in the terrestrial (left) and microgravity (right) environments.

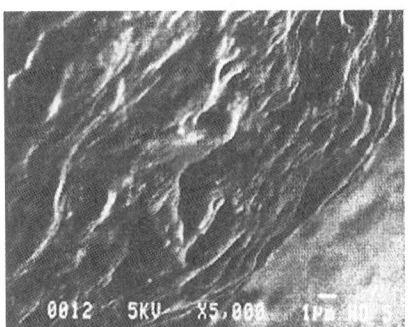
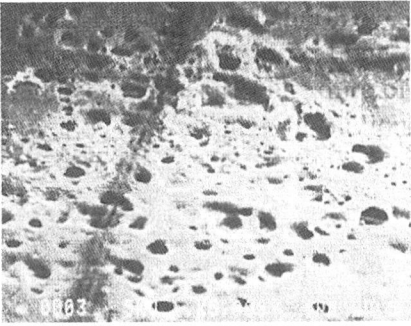

Figure 12. High-resolution SEM micrographs of the surfaces of the polymer networks shown in Figure 11. Terrestrial (left) and Microgravity (right). The scalebar in both images has a length of 1.0 micrometer.

Summary

We have produced PDLC and NSLC materials in microgravity and terrestrial environments using polymerization-induced phase separation (PIPS). The reduction in density difference between the liquid crystal and polymer results terrestrial samples with larger droplets and a higher number of droplets per unit volume than the corresponding microgravity samples for PDLCs with liquid crystal percentages (e.g. 45% in NOA65) close to the minimum required for phase separation. For high percentages of liquid crystal, the reduction in density difference achieved in microgravity results in a narrower droplet size distribution and more spherical droplets as compared to the terrestrial samples. The electro-optic properties are more desirable for the high liquid crystal (60%) samples processed in microgravity than the samples with other liquid crystal percentages. The NSLC samples exhibited differences in macroscopic alignment and in the texture of the polymer network itself. More extension terrestrial investigation is required to design microgravity experiments to fully elucidate the gravitational influence on the morphology of PDLC and NSLC materials.

Acknowledgments

We gratefully acknowledge the support of NASA (NAG13-28), Marshall Space Flight Center (KC-135 flight time), and NSF (EPSCoR and DMR-9512506).

References

1. Crawford, G.P.; Whitehead, J.B. Jr.; Zumer, S. In *Optical Properties of Polymer Dispersed Liquid Crystals*; Elston, S.; Sambles, R., Eds.; Taylor and Francis: London, 1998; pp 233 – 288.
2. Crawford, G. P.; Doane, J.W.; Zumer, S.; *Polymer Dispersed Liquid Crystals and Related Systems*; Oxford University Press: Oxford; 1997.
3. LeGrange, J. D.; Miller, T.M.; Wiltzius, P.; Amundson, K.R.; Boo, J.; Van Blaaderen, A.; Srinivasarao, M.; Kmetz, A. *SID Dig. Tech. Papers* **1995,** *XXVI*, 275.
4. Drzaic, P.S.; Gonzales, A.M.; Jones, P.; Montoya, W. *SID Dig. Tech. Papers* 1992, *XXIII*, 571.
5. Drziac, P.S. *Polymer Dispersions*; World Scientific: Singapore; 1996.
6. Nagae, Y.; Ando, K.; Asano, A.; Takemoto, I.; Havens, J.; Jones, P.; Reddy, D.; Tomita, A. *SID Dig. Tech. Papers* **1995,** *XXVI*, 223.

7. Lawandy, N.M.; Firehammer, J.A.; Vartak, S.D.; Crawford, G.P. *J. Soc. Info. Dis.* **1997**, *5*, 383.
8. McCargar, J.W.; Ondris-Crawford, R.J.; West, J.L. *J. Elec. Imag.* **1992**, *1*, 22.
9. Doane, J.W.; Vaz, N.A.; Wu, B.-G.; Zumer, S. *Appl. Phys. Lett.*, **1986**, *48*, 269-271.
10. Doane, J.W.; Yang, D.K.; Chien, L.C. Intl. Disp. Res. Conf. **1991**, IEEE, 175-178.
11. West, J.L. *Mol. Cryst. and Liq. Cryst.* **1988**, *157*, 247
12. *Liquid Crystals in Complex Geometries;* Crawford, G.P.; Zumer, S., Eds.; Taylor and Francis: London, 1996.
13. Serbutoviez, C.; Kloosterboer, J.G.; Boots, H.M.J.; Touwslager, F.J. *Macromol.* **1996**, *29*, 7690-7698.
14. Whitehead, J.B., Jr.; Zumer, S.; Doane, J.W. *J. Appl. Phys.* **1993**, *73*, 1057.
15. Wu, B.-G.; Erdmann, J.H.; Doane, J.W. *Liq. Cryst.* **1989**, *5*, 1453-1465.
16. Fox, B.H.; Schuster, P.A.; Kilp, T.; Fuh, A.Y.G. *AIAA* **1993**, 93-0578.
17. Jin, J.-M.; Parbhakar, K.; Dao, L.H. *Langmuir,* **1996**, *12*, 2096.
18. Jin, J.-M.; Parbhakar, K.; Dao, L.H. *Macromol.* **1995**, *8*, 7937.
19. Parbhakar, K.; Jin, J.-M.; Nguyen, H.M.; Dao, L.H. *Chem. Mat.* **1996**, *8*, 1210.
20. Parbhakar, K.; Dao, L.H. *Space Forum* **1998**, *3*, 277 – 306.
21. Whitehead, J.B.; Chandler, M.M. *Proc. SPIE* **1997**, *3123*, 128.

Chapter 11

Processing of Nonlinear Optical Polymer Systems under Microgravity Conditions

L. H. Dao, K. Parbhakar, H. M. Nguyen, J. M. Yin, Y. Sun, and Y. Beaudoin

Advanced Materials Research Laboratory, INRS-Energie et Matériaux, University of Québec, 1650 Boulevard Lionel-Boulet, Varennes, Québec J3X 1S2, Canada

We report here the results of our microgravity experiments which were carried out at the ZARM drop tower facility at Bremen, Germany. Two systems were tested: (a) guest-host system using thermal processing and (b) cross-linked system using photocross-linking processing. Five drops of 4.75 sec of microgravity were made and for each drop 15 guest-host samples and 9 photo cross-linking samples were processed. The sample thickness was fixed at 2 μm, however poling field, UV intensity, chromophore type and concentration, poling temperature and cooling rate were varied. The relaxation behavior was monitored using second harmonic generation as a function of time for both the samples prepared at 10^{-6} g and 1 g. Improvement of relaxation time was observed for certain samples. Simulations using a modified Robertson-Shimha-Curro (RSC) theory were also carried out. Qualitatively experimental results support the simulation hypothesis.

Introduction

With the rapid increase in the rate of information transmission via high-speed optical fibers and the need for increased storage density in optical storage devices, there is a strong need of commercial devices using nonlinear optical (NLO) polymers (*1*). Conventionally, inorganic crystals were used for optical applications such as waveguides, optical modulators, optical memory storage, and holography, because of their high nonlinear optical performance. However, long times and delicate controls are required to grow crystals in the large sizes and optimized molecular structures needed for NLO devices. Because of the versatility, ease of fabrication, and high optical grade of polymers, it has been shown that there are excellent potentials and benefits in developing nonlinear optical polymers (*1-4*). Poled-polymer systems (*3-5*) due to their large NLO coefficients, fast response time, intrinsic tailorability, and processibility for integrated optics, have been identified to be practical in electro optic devices. Several different molecular-design approaches for NLO polymers are known: (a) in guest-host systems, the NLO dipoles are simply dopants in the host material; (b) in side-chain polymers they are chemically bonded to the backbone polymer via flexible spacer units; (c) in main-chain polymers the NLO dipoles are part of the main polymer chain, and (d) in cross-linkable polymers they are chemically cross-linked with the backbone molecules under an applied electric field.

The successful implantation of poled polymer films depends largely on the outcome of research aimed in developing materials in which dipole orientation will be retained over an extended period of time in the operating temperature range. An important element of this research comprises not only the determination of process and material factors that affect stability of polar order but also the mechanisms involved in relaxation which translate into loss of optical second-order nonlinearity (χ^2) in the material. These issues have been addressed in a number of studies on guest-host, side-chain, and cross-linked polymers systems using electro optic (EO), second-harmonic generation (SHG), thermally stimulated current (TSC), and dielectric relaxation measurements (*4,5*). In several cases, films maintained at room temperature show negligible decay in the second-order nonlinear response over several weeks, but seldom exhibit adequate stability for more than a few days at temperatures approaching 100°C. Explanations for decay in χ^2 at elevated temperatures have been centered around oriental relaxation, using a number of different models which generally have employed bi-exponential expressions to describe the time constants. The temperature dependence of relaxation times has been analyzed by Arrhenius plots and Williams-Landel-Ferry expression.

The use of second-order nonlinear polymers in devices however still awaits solution to a number of key problems, e.g., why glass-transition temperature is the dominant parameter of relaxation time, why the nature of the chromophore or the polymer host does not play a more dominant role, how to simultaneous optimize poled-order stability and electrooptic coefficient?

Although several methods exist to generate NLO materials, our ability to tailor them for specific applications is still primitive, because we lack adequate structural data and models for understanding the poling process and the dipole relaxation, the relationship between the chromophore Brownian motion (orientation) and the mobility of the host polymer chains and the local free volume present in the vicinity of the chromophores. Measurement of poled-order decay can be done by following either the SHG, electro optic or birefringence. Attempts are made to relate the poled-order relaxation process to relaxation processes measured in polymers using other experimental techniques such as dielectric relaxation (6). Resonances in the dielectric relaxation spectrum of polymers are indicated by the term α, β, γ. α-Relaxations are usually identified with the glass transition and involve relative motions of polymer chains. β-Relaxtions are frequently associated with local motions of side groups on the polymer chain. Dielectric and thermally stimulated discharge (TSD) experiments have indicated that the rotational motion of the chromophore responsible for decay of poled order is related to the relaxation process which is in turn related to the glass transition temperature. The Vogel-Tamann-Fulcher (VTF) equation has proven useful in relating viscoelastic properties of polymers to the glass transition temperature at temperatures above T_g. Below Tg, it is generally assumed that the functional form of the time-temperature relationship becomes Arrhenius-like. The VTF equation is often explained using a free volume model (7). Free volume can be described reasonably precisely from a theoretical point of view as that portion of the total unoccupied volume in an amorphous system that can be redistributed without a change in the free energy.

In this study we are looking on the effect of the microgravity environment on the temporal stability of the poled thin NLO films. Experiments were carried out at the ZARM drop tower facility at Bremen Germany. Two systems were tested: (a) a Guest Host system was prepared using thermal processing and (b) cross-linked specimens were produced using photo cross-linking procedures.

Experimental

Drop capsule for ZARM Bremen Drop Tower.

The microgravity experiments were carried out at ZARM Drop Tower using a 7-platform drop capsule. From bottom to top, there are (1) platform No.

1: on-board batteries and electronics; (2) platform No. 2: on-board computer; (3) platform No. 3: large DC-AC converter and powersupply; (4) platform No. 4: photo processing system PPS; (5) platform No. 5: controllers for PPS and TPS; (6) platform No. 6: thermal processing system TPS and finally (7) platform No. 7: water cooling system (see references (8) and (9) for details).

The Photo Processing System (Platform No. 4) consists of (1) a large surface high power monochromic UV lamp (wavelength = 365 nm and power = 100 mW/cm^2) with a mechanical shutter activated by an electrical pulse and an AC stabilized power supply, (2) a cell support and (3) probes and controllers for control and data acquisition during the drop experiment.

Figure 1. Sample Holder with NLO devices

The ceramic cell support (Figure 1) can hold three sets of experiments under different control factors such as lamp intensity, transverse E-field, controlled heating (temperature). Each set consists of three identical cells in order to insure the reproducibility of the results. We have designed the PPS door system in order to approach the sample holder to about 1 cm of the UV lamp. With this design, it is possible to open the shutter 0.25 to zero second before the drop. Several probes and controllers are also fitted on this platform which are used for the control and the data acquisition during the drop.

The Thermal Processing System was used for the preparation of samples in the Guest Host system. Here again the ceramic sample holder was able to hold three sets of experiments, however each experiment was conducted on five samples. A powerful radiative heater was used to heat the samples above Tg. A water-cooling heat exchanger was necessary to rapidly cool the samples using a fan. A DC field was used for electrical poling and each sample was fitted with a fast acting fuse which eliminated the shorted samples only during the drop experiment. Several temperature controllers were used to measure the temperature distribution on the sample holder. The heating and cooling was performed automatically.

NLO Devices Fabrication

A piece of ITO glass (2cm x 3.5cm) was covered with Teflon tape on its ITO surface. Some part of the Teflon tape was cut and peeled off. Then the ITO glass covered with the Teflon tape was immersed into concentrated hydrochloride acid for 8-10 min to allow the ITO without the protection of the Teflon tape to be etched off. Afterwards, the Teflon tape was peeled off and the ITO glass was washed with water, acetone, chloroform successively and dried in vacuum at room temperature for several hours before it was ready for spin coating.The solution of chromophore, polymer and organic solvent was filtered through the 0.5µ filter before spin coating. The thickness of the spin coated film is dictated by both the spin speed and the viscosity of the solution. For example, in order to obtain the Disperse Red 1(DR1)/polymethyl methacrylate (PMMA) spin coating film, a solution of about 0.1 g/ml PMMA in chloroform with suitable amount of DR1 was filtered and spin coated onto the ITO side of ITO glass treated above. The film was dried at 50°C in vacuum for several hours and ready for used. DR1 side-chain polyimide and polycarbonate-chromophore films were also obtained in the same way.

The NLO device was obtained when two pieces of ITO glass were placed in parallel face to face (the two coated polymer films contacted each other) with the overlapped area about 2.5cm x 2 cm, and pressed with Carver Laboratory Press instrument at 140°C and about 100-300 lbs pressure for about 10 min to obtain the ITO glass -sample- ITO glass sandwich with the window area about 1.0cm x 1.5cm. The two ITO plates will serve as electrodes. The finished samples are shown in Figure 1. In the case of photo-crosslinking samples, a cavity which was created between two ITO plates by a polyimide spacer, was filled with a heated blend of DR1 or c-TBA and polyvinylstyrylacrylate (PVSA). The cross-linking was accomplished by exposing the heated viscous solution of PVSA/DR1 or c-TBA to UV radiation (λ = 365 nm).

Characterization (Relaxation of Second Harmonic Signal)

Because of their application as electro-optic (EO) materials the appropriate choice is to measure Pockels coefficients v_{ij} which alter the propagating characteristics of light in the medium. However, we found v_{ij} too small for any quantitative analysis of the microgravity effects, instead, we opted to monitor the relaxation of second harmonic (SH) signal.

Results and Discussion

Free Volume and Simulations

Free-volume theory is an improved cell or lattice model for the liquid state by introduction of vacancies in the lattice. In the free-volume theory, the Simha-Somcynsky equation of state of a polymer system is written as (*10*)

$$\frac{\underline{P}\,\underline{V}}{\underline{T}} = [1 - y(2^{\frac{1}{2}}y\underline{V})^{\frac{1}{3}}]^{-1} + \frac{y}{\underline{T}}[2.022(y\underline{V})^{-4} - 2.409(y\underline{V})^{-2}] \quad (1)$$

where $\underline{P}$, $\underline{V}$ and $\underline{T}$ are reduced variables, $\underline{P}=P/P^*$, $\underline{V}=V/V^*$, and $\underline{T}=T/T^*$, and P^*, V^* and T^* are the characteristic scaling parameters for each polymer material and can be deduced from experimental data, y is the fractional occupancy of the cells and the fractional free volume f is defined as

$$f = 1 - y \quad (2)$$

In practical calculations, a polymer sample is usually divided into smaller regions, each region contains N_s monomer segments (the region is therefore denoted as N_s-size region) and an amount of free volume which is assumed to be a multiple of a quantity denoted by β. The free volume in these regions is then one of the amounts, 0, β, 2β, ..., (j-1)β, ..., (n-1)β. Here n is the total number of states. j runs from 1 to n and describes the set of free-volume states. The free volume in the system is usually assumed to take a binomial distribution, i.e.,

$$\xi_j = \binom{n-1}{j-1} p_r^{j-1}(1-p_r)^{n-j} \quad (j=1,2,\ldots,n) \quad (3)$$

where ξ_j (called state occupancy) is the relative probability that a given N_s-size region has the free volume (j-1)β. The parameters β and p_r are determined by the following equations,

$$\begin{cases} (n-1)p_R\beta = f_{av} \\ (n-1)p_r(1-p_r)b^2 = \langle \delta f^2 \rangle_{av} \end{cases} \quad (4)$$

where f_{av} and $\langle \delta f^2 \rangle_{av}$ are the mean free volume and the root-mean-square fluctuations in free volume. The occupancies sum to unity, i.e.,

$$\sum_{j=1}^{n} \xi_j = 1 \tag{5}$$

To describe the kinetics of the structure relaxation process following sudden cooling or heating, the relaxation time τ is first of all empirically related to the computed Simha-Somcynsky free volume f for the equilibrium liquid at a constant pressure, i.e.,

$$\tau^{-1}(T) = (\tau_g a_T)^{-1} + (\tau_b b_T)^{-1} \tag{6}$$

where τ_g is the nominal relaxation time at T_g and τ_b is the relaxation time of sub-T_g motion at T_g.

The parameters a_T and b_T are determined by the following equations,

$$\log a_T = -c_1 + \frac{c_1 c_2}{c_2 + (f - f_g)T^*/f^*} \tag{7}$$

$$\log b_T = -C_1 + \frac{C_1 C_2}{C_2 + (f - f_g)T^*/f^*} \tag{8}$$

where c_1, c_2, C_1 and C_2 are the time-temperature shift parameters, f_g is the free volume at the glass transition point, and f^* is called the characteristic free volume. A further assumption (*11,12*) is that the above equations apply to any local regions (which are assumed to attain local equilibrium), whether or not the whole system is in equilibrium. Now considering all possible distinct N_s-size regions in a polymer material, the fraction of these which at time t have the free volume (j-1)β is denoted by $w_j(t)$. Then the equation governing the change in $w_j(t)$ from an initial time t_0 to present time t is written as (*13*):

$$w_j(t) = \sum_{i=1}^{n} w_i(t_0) P_{ij}(t - t_0) \quad (j = 1, 2, \ldots, n) \tag{9}$$

where $P_{ij}(t-t_0)$ is the transformation probability that the free-volume state in a given region changes from i to j in the time interval $(t-t_0)$. And $P_{ij}(t-t_0)$ satisfies the following equation,

$$\frac{dP_{ij}(t)}{dt} = \sum_{k=1}^{n} P_{ik}(t) A_{kj} \tag{10}$$

where matrix A is tridiagonal, its components are written as

$$\begin{cases} A_{k,k+1} = \lambda_k^+ & (k=1,2,...,n-1) \\ A_{k,k-1} = \lambda_k^- & (k=2,3,...,n) \\ A_{kk} = -(\lambda_k^+ + \lambda_k^-) & (k=1,2,...,n) \\ A_{kj} = 0 & (j \neq k-1, k, k+1) \end{cases} \tag{11}$$

with λ_k^+ and λ_k^- being the upward and downward transition rates from state k, respectively. At equilibrium, the state occupancies ξ_k, are assumed to be unchanging, and the detailed balancing condition leads to the following,

$$\xi_k \lambda_k^+ = \xi_{k+1} \lambda_{k+1}^- \quad (k=1,2,...,n-1) \tag{12}$$

The individual upward and downward rates are expected to be proportional to the relaxation rate τ^{-1} which is given by Eq. (6), i.e.,

$$\begin{cases} \lambda_k^- = \left(\frac{\xi_{k-1}}{\xi_k}\right)^{1/2} R\beta^{-2}\tau^{-1} \\ \lambda_k^- = \left(\frac{\xi_{k-1}}{\xi_k}\right) \lambda_{k+1}^- \end{cases} \tag{13}$$

where R is a constant.

Eq.(10) can be solved by standard matrix diagonalization procedure. Once the eigenvalue problem of the matrix A is solved, $w_j(t)$ and the average free volume f (as well as the value of y in Eq.(2)) are known, the volume V of the system can then be obtained from Eq. (1). Therefore, the relaxation process is fully determined (RSC model).

Gravity is an important parameter which affects the relaxation process following sudden cooling. To take the gravity into account in the RSC model, we further assume that the downward transition rate λ_k^- is proportional to the free volume in a given region, i.e., $\lambda_k^- \propto (1+\alpha_G k)$. Here k is the index of the state of occupancy and α_G specifies the strength of the gravity field. That is, the larger the size of a free space, the larger the downward transition rate λ_k^- is. Thus λ_k^- can be written as

$$\lambda_k^- = (\frac{\xi_{k-1}}{\xi_k})^{1-2} R \beta^{-2} \tau^{-1}(1+\alpha_G k) \tag{14}$$

The physical reason behind this assumption is as follows: during the relaxation process, the polymer molecules relax into their nearby free spaces (or voids), and the gravity force plays a role as to enhance such relaxation by exerting an additional force on the polymer molecules.

Starting from some initial equilibrium state I (V_i, T_i), a polymer (e.g., poly (vinyl acetate)) is suddenly cooled down to a lower temperature T_f. We examine in which way the sample relaxes to its equilibrium state F (V_f, T_f). When the sample is suddenly cooled to T_f, the volume of the sample will usually not assume the value of the final state F, rather it will first go through some intermediate state M (V_m, T_f). Here V_m is normally not the same as V_f and can be determined by $V = V_g[1+\alpha_g (T-T_g)]$, where V_g is the volume at the glass-transition temperature T_g and α_g the thermal expansion coefficient of the glass state. Since the state M is not an equilibrium state, and therefore, it will relax towards its equilibrium state F. Our interest here is to examine the kinetics involved in the relaxation process and investigate how gravity affects the relaxation process.

Simulations using free volume theories were performed where the gravitational parameter αg was arbitrarily varied. Lower αg represents reduced gravity environment. Figure 2 shows the volume relaxation of poly (vinyl acetate) after sudden cooling from an initial $T_i = 313$ K to final $T_i = 303$ K for three different values of α_g, where the curves A, B and C correspond to $\alpha g = 0.5$ (A), 5.0 (B) and 15.0 (C) respectively and $v_f =$ is the final volume at $T_f = 303$ K for $\alpha_g = 0$. We note that simulation results are qualitatively in agreement with experiments. An exact comparaison is difficult at this time because the simulation parameters for DR1 and PMMA are not available in the literature and must be measured.

The main observation for the sudden cooling is that the gravity accelerates the structure relaxation process. As said earlier, this is because of the fact that the gravity force increases the chance for the polymer molecules to relax into the free space in the polymer material. The relaxation time t_R which is required for the system to reach its equilibrium state after a sudden change in temperature decreases with the gravity parameter α_g, as shown in Figure 2.

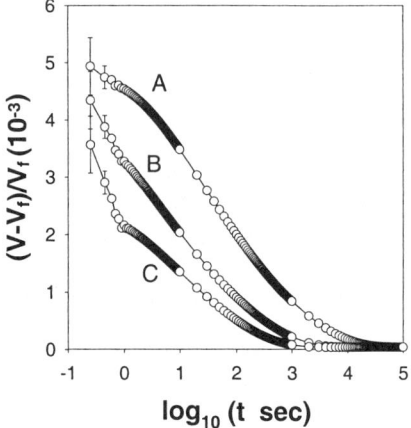

Figure 2. Volume recovery process of poly (vinyl acetate) after a sudden cooling from $T_i = 313$ K to $T_f = 303$ K for different gravity parameter α_g where curves A, B and C are for $\alpha_g = 0.5$ (A), 5.0 (B) and 15.0 (C), respectively; and V_f is the final volume at $T_f = 303$ K for $\alpha_g = 0$

To study the effect of cross-linking reaction on relaxation process, the transition rates are modified. We note that τ of many glass-forming liquids has an identical form as that of the viscosity η. The η of polymer system increases exponentially with time t near the gel point, i.e., $\eta \propto \exp[\alpha_p t]$ with α_p being the cross-linking reaction rate. Therefore we assume that $\tau \propto \exp[\alpha_p t]$, as a result, λ_k^- can be modified as

$$\lambda_k^- = (\frac{\xi_{k-1}}{\xi_k})^{1-2} R \beta^{-2} \tau^{-1}(-\alpha_p t) \tag{15}$$

Figure 3 shows the volume recovery process after a sudden cooling for three different values of α_p where the initial and final temperature are, respectively, $T_i = 313$ K and $T_f = 303$ K. The curves A, B and C are for $\alpha_p = 2 \times 10^{-3}$, 4×10^{-3} and 8×10^{-3}, respectively. Basically from Figure 3 we note that the curves show similar variation except that the final constant value of V depends on α_p, and with larger α_p we get a higher value of V implying a slower relaxation. For the range of α_p studied in our analysis V can be enhanced by almost 50 %.

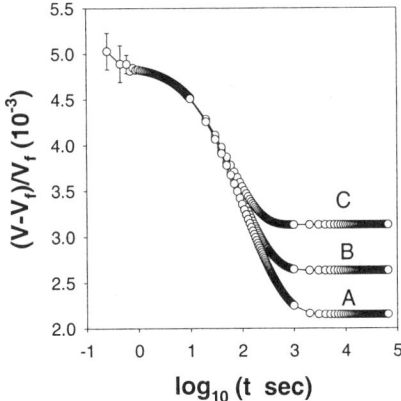

Figure 3. Volume recovery process of poly (vinyl acetate) after a sudden cooling from $T_i = 313$ K to $T_f = 303$ K for different cross-linking rate parameter α_p, where curves A, B and C are for $\alpha_p = 0.5$ (A), 5.0 (B) and 15.0 (C), respectively; and V_f is the final volume at $T_f = 303$ K for $\alpha_p = 0$

Experimental Results

Experiments were carried out at the ZARM drop tower facility at Bremen, Germany. Five drops of 4.75 sec of microgravity were made and each experiment was repeated at 1 g right after the drop using similar set of samples and parameters. Several combinations of chromophore concentration, type of chromophore and polymer, poling temperature and applied voltage, sample thickness, etc., were studied.

Table 1 gives the layout of specimens and the typical parameters used in the thermal NLO (TNLO) processing of DR1/PMMA films. Although the initial heating temperature was fixed at 100 °C, the real temperatures of each sample were variable and different from the center (A2 and A3) to the side (A1 and A4). The cooling rates of Column C were double of Column A and the initial and final temperatures after 4.75 sec. were also different for each sample depending on the position. Similar arrangement was used for the photo cross-linked NLO (PNLO) processing of films except that in this case we only used three samples in each column. Due to the short microgravity time, the cross-linking reaction was not completed after 4.75 sec and in fact continued during the capsule recovery time at 1 g. Therefore, no data are presented for the Photo Process System. The SH signal was measured using the procedure described above. In general, we observed a considerable improvement in the SHG signal in the samples prepared at 10^{-6} g as compared to the samples prepared at 1 g. So far we have measured the SH signal intensity at three points in time.

TABLE 1

Colum A Specimen	Thickness μ	T(°C)	V
A1 DR1(5%)+PMMA	2	100	30
A2 DR1(5%)+PMMA	2	100	30
A3 DR1(5%)+PMMA	2	100	30
A4 DR1(5%)+PMMA	2	100	30
Column C Specimen	Thickness μ	T(°C)	V
C1 DR1(5%)+PMMA	2	100	30
C2 DR1(5%)+PMMA	2	100	30
C3 DR1(5%)+PMMA	2	100	30
C4 DR1(5%)+PMMA	2	100	30

Table 2 and Table 3 give, respectively, the SHG intensity for the samples processed at 10^{-6} g and at 1 g, and measured after 10 days, 120 days and 150 days after their processings. The SH signal intensity was measured on two sides of the specimen to find out the polarization direction. The side which gave the higher signal was the correct side and the subsequent measurements were made for that side at three different points.

We note that there is considerable enhancement of SH signal for samples prepared under Microgravity (Table 2) compared to the same samples prepared at 1 g (Table 3). This finding is interesting but it is to early to draw any definite conclusions on the effect of Microgravity on the devices properties and the aging process. Experiments need to be repeated before we can make any definite claims.

TABLE 2. SHG signal (mV/V) at 10^{-6} g

Sample	10 days	120 days	150 days
A1	140	113	90
A2	170	73	76
A3	330	165	117
A4	230	156	115
C1	680	500	280
C2	460	335	250
C3	320	109	80
C4	1050	985	690

TABLE 3. SHG signal (mV/V) at 1 g

Sample	10 days	120 days	150 days
A1	100	70	-
A2	100	80	-
A3	130	65	39
A4	190	95	43
C1	190	125	59
C2	170	94	55
C3	190	103	60
C4	100	98	46

Conclusions

We have used ZARM drop tower for the first time to fabricate NLO polymer films using thermal and cross-linking polymerization processes. Several combinations of the parameters, which control the characteristics of the NLO materials, have been used under microgravity environment. The experiments are repeated at 1 g under similar conditions. The generation of SH signal is used to compare the poling order produced in the two samples which are prepared at 1 g and 10^{-6} g. The aging process is also studied by monitoring the SH signal as a function of time. The preliminary results are very optimistic in that the samples prepared under 10^{-6} g show higher polar ordering. This result is very interesting, however we do not have sufficient data to come to definite conclusions.

Acknowledgments

This work was supported by the Canadian Space Agency (CSA), Microgravity Sciences Program, the Natural Sciences and Engineering Research Council of Canada (NSERC) and ZARM (Germany). The authors wish to express deep appreciation to Dr. P. Gregory, program manager, Dr. R. Herring, program scientist and Mr. S. Desjardins, project manager of CSA for their constant interest and support.

Literature Cited

1. *Nonlinear Optical Properties of Organic and Polymeric Materials;* Williams, D.J., Editor; ACS Symp. Ser., Vol. 233 (1994), American Chemical Society, Washington, DC.
2. *Nonlinear Optical Properties of Organic Materials and Crystals;* Chemla D., Zyss J., Eds., Academic Press, 1987
3. *Nonlinear Optical Effects in Molecules and Polymers;* Prasad, D. W.; John Wiley and Sons, N.Y., 1991
4. Burland D.M., Miller R.D., Walsh C.A. *Chem. Rev.* **1994**, *94*, 31
5. *Quantum Electronics: A Treatise;* Rabin H., Tang C.L., Eds., Academic Press, N.Y. 1975
6. *Dielectric Spectroscopy of Polymers;* Hedvig P., John Wiley and Sons, N.Y., 1977
7. Turnbull D., Cohen M.J. *J. Chem. Phys.* **1961**, *34*, 120
8. Jin J.M., Parbhakar K., Dao L.H. *Macromolecules* **1995**, *28*, 7937
9. Parbhakar K., Jin J., Nguyen H., Dao L.H. *Chem. Materials* **1996**, *8*, 1210
10. (a) Simha, R.; Somcynsky, T. *Macromolecules* 1969, 2, 342. (b) Simha, R. *Macromolecules* **1977**, *10*, 1025.
11. (a) Robertson, R.E.; Simha, R.; Curro, J.G. *Macromolecules* **1984**, *17*, 911. (b) Robertson, R.E.; Simha, R.; Curro, J.G. *Macromolecules* **1985**, *18*, 2239. (c) Simha, R.; Curro, J.G.; Robertson, R.E. *Polym. Eng. Sci.* **1984**, *24*, 1071. (d) Robertson, R.E. *Macromolecules* **1985**, *18*, 953. (e) Robertson, R.E.; Simha. R.; Curro, J.G. *Macromolecules* **1988**, *21*, 3216.
12. Robertson, R.E. in *Computational Modelling of Polymers* (Bicerano J. ed.); Marcel Dekker Inc.: New York, 1992; p.297.
13. van Kampen, N.G. *Stochastic Processes in Physics and Chemistry;* North-Holland: Amsterdam, 1981.

Ground-Based Research

Chapter 12

Foaming of Polyethylenes in a Dynamic Decompression and Cooling Process

Kwangjin Song and Robert E. Apfel

Department of Mechanical Engineering, Yale University, 9 Hillhouse Avenue, New Haven, CT 06520

Solutions of polyethylene melt and blowing liquid have been foamed in a dynamic decompression and cooling (DDC) process to fabricate lightweight polymers. The cellular structures of resultant foams are open or closed, with micromorphologies of granules, fibers, and fiber-networks. The foamed polymers contain anisotropic cells and crystals which increase mechanical strength. It is expected that in microgravity, DDC foaming will be able to produce uniform polymer solutions, resulting in foamed structures with enhanced properties.

Polymer foams can be generated by a number of technologies (*1*). Among them, the expansion process is the most widely used technique wherein gas bubbles nucleate and expand either by decompression or heating of the system that contains polymers and blowing agents. Its foaming mechanism generally involves the pressure difference between the inside of the cell and the surrounding medium. The expansion process driven by decompression that uses gases as a blowing agent was introduced in the early 1930s by Munters and Tandberg in their pioneering polystyrene (PS) foaming patent (*2*). In 1950, this batch technique was modified into a continuous process by McIntire of the Dow Co. (*3*). Rubens et al., also from the Dow Co., proposed in 1962 another notable process that employs 1,2-dichlorotetrafluoethane as a foaming gas for polyethylene (PE) (*4*). Simultaneously in the 1960s, a technology that uses liquids as a blowing agent was patented by Blades and White of the DuPont Co.

(5,6). Here, homogeneous mixtures of polymer/poor solvents are decompressed through a shaping die into a closed-cellular body.

A DDC process, developed in 1995 by Apfel, was directed at creating bulk, open-cell foams of glassy metals using an immiscible pair of melt and blowing liquid (7). In the process, a melt seeded with droplets of a volatile liquid under non-condensable gas pressure is rapidly decompressed leading to a dynamic, superheated and supersaturated mixture; then the droplets explosively vaporize, homogeneously cooling and expanding the melt. Depending on the cooling rate of the melt, either amorphous or crystalline foams can be manufactured. For polymer foaming, we employed poor solvents as a blowing liquid, following the previous method of Blades and White (5,6), and then produced the open cell foams of poly(butylene terephthalate, PBT) (8). The DDC-foamed polymers have complex features both in macrostructure and microstructure that may, in some instances, follow a phase separation and a fractal growth mechanism.

Background

DDC-Induced Supercooling

Consider a DDC system that contains a binary liquid solution of polymer melt and blowing liquid. As the decompression is effected above the boiling point of the solvent, the volatile solvent phase vaporizes to cool the melt by taking latent heat from it. Assuming that the two species are non-reacting and form a homogeneous binary phase, one can write an energy equation:

$$\Delta T(\rho_m C_{p,m} V_m) = \Delta H_{v,s} \rho_s V_s - (\Delta H_{f,m} + \Delta H_{c,m})\rho_m V_m \qquad (1)$$

where, ΔT is the reduction in temperature; the subscripts m and s denote respectively the polymer melt and the solvent; ρ is the density; C_p is the specific heat; V is the volume; ΔH_v is the latent heat of vaporization; ΔH_f is the heat of fusion; and ΔH_c is the heat of crystallization.

In the DDC process, like other foaming technologies, the cooling rate of the polymer melts ($dT/dt \approx \Delta T/\Delta t$) is of practical importance in designing micro- and macroscopic structures of resultant foams. Bubble growth is related to vaporization of the solvent and hence to the rate of melt cooling, in which material properties such as density (ρ), diffusivity (D), viscosity (η), and surface tension (γ), etc. are functions of transient temperature and pressure.

A growth model of a large number of spherical bubbles in polymer melts was proposed, which employs a cell model and assumes an ideal gas and a Newtonian liquid (9). The bubbles are close in proximity and have a uniform radius R. This model has been modified to incorporate both non-Newtonian behavior of polymer melts and distributions of bubble sizes (*10-12*). The bubble growth models are however similar in principle and neglect the latent heat of the volatile phase that does not behave as a perfect gas. Continuity and momentum balances yield (*9*):

$$\frac{dR}{dt} = \frac{V+R^3}{4\eta V}\left[(P_g - P_f)R - 2\gamma\right] - \frac{R}{3}\frac{d\ln\rho}{dt} \qquad (2)$$

$$\frac{\partial c}{\partial t} = 9D\rho^2 \frac{\partial}{\partial x}\left[\left(\frac{x}{\rho} + R^3\right)^{4/3}\frac{\partial c}{\partial x}\right] \qquad (3)$$

where, a convected coordinate, $x = (r^3 - R^3)\rho$; V is the volume of a cell; P_g and P_f are the pressures respectively inside the gas bubble and in bulk liquid surrounding a cell; γ is the surface tension; and c is the concentration of dissolved gas in the solution. The density term in Eq. (2) will dominate in bubble growth as the melt solidifies, i.e., $\eta \rightarrow \infty$, or as the system approaches mechanical equilibrium. In the limit, the final bubble size (R_f) is related to the initial bubble size (R_i) through $R_f = R_i\,(\rho_f/\rho_i)^{-1/3}$.

Effects of Gravity on Foaming

The rate of bubble nucleation (J_o) in a liquid phase depends greatly on degree of supersaturation and the surface tension (γ) of the gas-liquid interface. At equilibrium, the nucleation rate is written (*13*):

$$J_o = A\,exp\left(\frac{-\Delta G^*}{k_B T}\right) \qquad (4)$$

where A is the prefactor; the free energy of formation of the critical nucleus, $\Delta G^* = 16\pi\gamma^3/3\Delta P^2$ for homogeneous nucleation, k_B is the Boltzman constant, and T is the temperature. The pressure difference (ΔP) between the gas bubble and the surrounding medium, $\Delta P = 2\gamma/r$ and for two bubbles having radii r_1 and r_2, $\Delta P_{1,2} = 2\gamma(1/r_1 - 1/r_2)$. It is implied that lowering surface tension facilitates bubble nucleation and reduces the pressure difference between bubbles, leading to small average cell sizes.

Solutions containing a polymer melt, a poor solvent, and nitrogen gas in the DDC process tend to phase separate, the extent of which depends greatly on density differences and intermolecular interactions. The buoyancy force that results causes bubbles formed by decompression to ascend, though the effect is limited by the viscoelastic property of polymer melts, thereby imposing directional features on diffusion of gas phases. Also, gravity drains the melt out of cell walls and tends to collapse liquid foams.

A reduced gravity environment may influence polymer foaming in various ways. In particular, surface tension, which is normally masked by gravity, can dominantly control solution behaviors in microgravity; thus, the kinetics of nucleation and growth and the interactions between molecules could be altered. Due to the absence of buoyancy driven convection, molecular diffusion and Marangoni flow would dominate transportation of mass/heat in a quiescent state, thereby modifying the rate of mixing. Shear mixing, however, can better homogenize solutions without phase separation. It seems therefore that foaming under reduced gravity could generate lightweight polymers with cells and crystals smaller in size and narrower in dispersion than those prepared at equivalent conditions in unit gravity, a structural character that will probably increase a weight-to-strength ratio of resultant foams. It is expected, however, that DDC foaming in microgravity may not create large changes in bubble kinetics and dynamics because of its rapid decompression. Rather, a reduced gravity environment will provide suitable mixing conditions prior to decompression, particularly for polymer blends and composites such as polymers and metals.

Experimental

Materials and Foam Formation

The polymers used in this study were low- and high-density polyethylenes (LDPE, HDPE) with different melt indices (MI, g/10 min). The LDPEs with MI = 0.22 and 25 were from the Dow Chemical Co., and the LDPE with MI = 2.3 and the HDPE with MI = 0.5 were from the Mobile Chemical Co. HPLC-grades of chloroform ($CHCl_3$) and methylene chloride (CH_2Cl_2) with purity above 99.9 % of the Fisher Scientific Co. were used as blowing liquids.

Polymers in a solid pellet form were placed inside a high pressure vessel with varying concentrations in blowing liquids, and then pressurized with nitrogen gas (N_2) in the range of 2.07 to 6.21 MPa, i.e., above the vapor pressure of the solvent. While stirring at 10 rpm, the system was heated by an internal heater fitted with a glass crucible that contained the mixtures, to 160°C, above the crystalline melting temperature (T_m) of the polymers (9). The one-phase solution thus formed was decompressed at 110°C by quickly opening a release valve. Decompression was also conducted at 140°C by discharging the mixture at a constant pressure through a drain value into the atmosphere. Here the vessel employed an external heater. For comparison, a blown HDPE film with a blow-up ratio (BR) of one and a draw-down ratio (DR) of 20 was prepared by a Killon tubular film extruder equipped with an annular die.

Foam Characterization

The mean bulk densities of foams were determined by measuring the weight and volume of specimens that had a length of 50 mm prepared by freeze-cutting. The diameters were measured by a video camera. The volumes of some samples were estimated by a water displacement method.

Thermal analyses were performed by a Perkin-Elmer Pyris-1 differential scanning calorimeter (DSC). The samples weighing 10.0 ± 0.2 mg were scanned with a heating and a cooling rate of 20°C/min. The crystallinity (X_c) of samples was computed through X_c (%) = $(\Delta H_{exp}/\Delta H°) \times 100$, where $\Delta H°$ is the heat of fusion of 100% crystalline polyethylene (14).

Morphologies of the foams were observed with a Philips XL30 scanning electron microscope (SEM). Fresh surfaces of the samples parallel to the flow direction (FD) were prepared by freeze-fracturing or freeze-cutting using liquid nitrogen. The fresh surfaces were coated with a thin layer of carbon to prevent charging by the electron beam.

Wide angle X-ray diffraction (WAXS) patterns were taken using a Brucker GADD X-ray diffractometer. The middle part of the cylindrical foams was cut out along the FD and then divided into two specimens: the skin and core parts. WAXS patterns were taken through the normal direction (ND) and pole figures were constructed around two Eulerian angles at intervals χ of 5° and ϕ of 10°. Crystalline orientation factors were then calculated through (15):

$$f^B_{FD,j} = 2\overline{\cos^2 \phi_{FD,j}} + \overline{\cos^2 \phi_{HD,j}} - 1 \qquad (5a)$$

$$f^B_{HD,j} = 2\overline{\cos^2 \phi_{HD,j}} + \overline{\cos^2 \phi_{FD,j}} - 1 \qquad (5b)$$

where $\phi_{i,j}$ is the angle between the sample direction i and the crystallographic axis j. Here HD is the hoop direction of the foam.

An Instron Tensometer Model 4204 performed tensile testing at room temperature. The span length and the rate of stretching were respectively 15 mm and 10 mm/min. The foams were cut into 10 x 30 mm along the FD. The measured results were averaged from five samples tested at each condition.

Results and Discussion

Supercooling by DDC Foaming

In the DDC process, the reduction in temperature of the system depends not only on the properties but also on the concentration of the polymer/solvent mixtures, because the polymer melts supercool primarily by the latent heat of the evaporating solvent phase. The values of temperature reduction have been computed using Eq.(1) for the LDPE solutions (Figure 1). It was assumed that during decompression, the process of thermal equilibrium happens so fast throughout the system that the temperature in the melt volume is uniform. The $\Delta H_{f,m}$ of 35.7 cal/g for the highly crystalline DDC foams and the $C_{p,m}$ of 0.6122 cal/K·g at 110°C were taken (16).

The predicted results agree in general with the experiment, showing that solvents with higher latent heat of evaporation and solutions with lower concentration of the polymers yield deeper supercooling. The DDC process produced fairly high degrees of supercooling, enough to solidify the polymer melt in foaming. The two values, however, revealed a disparity that may arise from parameters such as a poor heat conductivity of the polymer, temperature of the surrounding medium as well as uncertainties in material data. Experiments also exhibited that the resins with lower MI tended to cool more slowly, narrowing the processing windows of the foaming process.

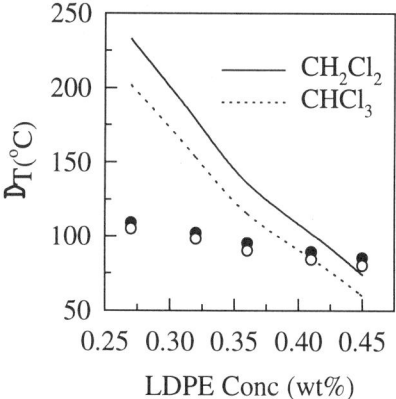

Figure 1. Temperature reduction (ΔT) for the LDPE solutions decompressed at 110°C; Line: Predicted values, Symbol: Experimental results (Filled: CH_2Cl_2, Open: $CHCl_3$).

Cellular Structure of DDC Foams

The DDC foaming that comprised the material inside the vessel produced a mixed foam structure of open and closed cells with average cell diameters varying from 50 to 1000 μm (Figure 2). In the core parts, more cells were open; however, more cells were closed in the skin parts, presumably because of the nonuniform distributions of the expansion stress in a glass container. Foam cells tended to elongate along the flow direction of the volatile solvent phase. A decrease of MI stabilized bubbles better. The LDPE foams with MI = 2.3 had smaller cells than those with MI = 0.22, and at MI = 25, foam cells were largely open.

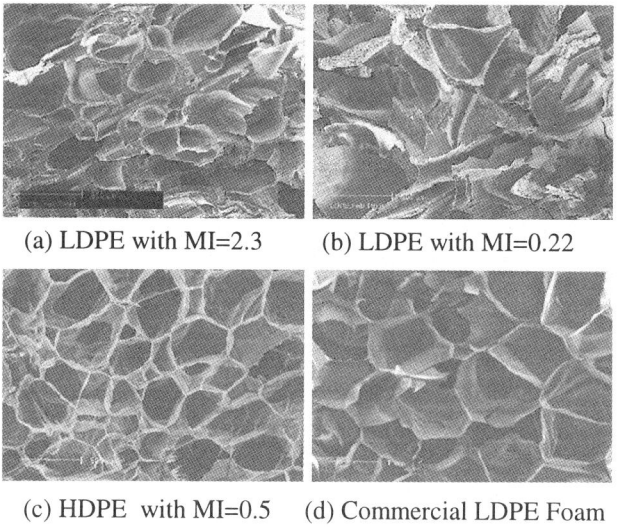

(a) LDPE with MI=2.3   (b) LDPE with MI=0.22

(c) HDPE with MI=0.5   (d) Commercial LDPE Foam

Figure 2. SEM micrographs of the DDC foams; (a)(b): The skin parts of the LDPE foams with $\rho_f=100$ kg/cm^3, (c) HDPE foam $\rho_f=10$ kg/cm^3, (d) Commercial LDPE foam with $\rho_f=40$ kg/cm^3.

The formation of open and mixed cell structures in the DDC process may be related in a complex way to many processing parameters. As cells grow in the melt, the cell walls become thinner with the melt draining out due to gravity and the capillary effect. A reduced gravity environment may better stabilize foam cells. The thinned walls could readily rupture if the expansion stress exceeds their mechanical strengths. However, the LDPE polymers exhibit self-stabilizing. In particular, the high viscosity of LDPE restricts the melt drainage and increases resistance of cell walls to thinning, thus stabilizing bubbles. Also, LDPE yields higher degrees of strain-induced hardening as compared with HDPE, owing to its branched molecular mechanism. It seems, therefore, that the formation of mixed cells involves more the foaming process *per se*, with explosively expanding bubbles and retarded cooling and crystallization of the melt, contributing to destabilization of bubbles. Parameters such as localized liquid/liquid phase separation and varying rates of nucleation and growth for individual bubbles could also be responsible. Bubbles nucleated in the later stage or grown slowly may remain closed without cell failure.

Foaming by discharging the melt into the atmosphere, on the other hand, produced closed cells with a mean cell diameter of ca. 500 µm. The foam cells were polyhedral in shape and quite uniform in size, which is comparable to a commercial foam with a mean cell size of ca. 1.0 mm. But, the DDC foams contained slightly collapsed cellular structures. Seemingly, this foaming process created a more rapid and uniform cooling of the melt, causing a rapid build-up of viscosity, which substantially stabilized bubbles.

The cellular morphology of closed cells depends on factors which determine the dispersion and shape of cells, including properties of foaming mixtures, gravity and the way foams are produced. In the DDC process, bubbles grow by the diffusion of N_2 and/or the volatile liquid phase. Due to the pressure differences between them, foam cells either grow or shrink. They also become dispersed by gravity-induced coalescence. As foam cells grow, they tend to be polyhedral by interactions. In polymer foaming, however, as the viscoelastic force plays a profound role in shaping, the structure of closed cells can greatly deviate from that at equilibrium (*17*). Thus, foaming in microgravity would cause changes in cellular structure due to differences in bubble dynamics and transport mechanism. McManus (*18*) found that polyurethane foams produced in microgravity possess much smaller and more uniform cells with more rounded shapes than those prepared in unit gravity. Pojman (*19*) reported that during frontal polymerization of benzyl acrylate and triethylene glycol dimethacrylate in microgravity, formed bubbles moved toward hotter regions by surface tension induced convection, thereby aggregating with each other. It is expected, however, that the changes of cell structures caused by reduced gravity may not be substantial in the DDC process since the polymer melts undergo rapid solidification during the transient decompression period.

Foam Density as a Function of Process Conditions

The mixed cell foams exhibited mean densities that fluctuated considerably depending on the position of the specimen. The skin and upper parts of the foams had higher densities than the core and bottom parts. Again, this local variation of the densities is probably due to the restriction of cell expansion in a glass container. At high expansion rates above 1.5 MPa/s, the mean densities of the foam decreased with increases in the decompression pressure (P_d) and temperature (T_d), but the changes were not substantial (Figure 3). At constant P_d and T_d, however, the mean foam densities increased appreciably, 90 to 160 kg/m^3, with decreases in the decompression rate (r_d). It seems that a retarded decompression suppresses the explosive evaporation of the volatile phase; the gaseous molecules may readily escape from the melt at relatively high temperatures, without causing a large degree of expansion.

The closed cell foams were uniform in shape, with much lower mean densities, below 10.0 kg/m^3, than those of the mixed cell foams, which tended to decrease with increases in the decompression pressure. As the melt was discharged into the atmosphere, the resulting foam experienced free expansion. However, at higher P_d, a sudden expansion developed large foam cells which subsequently collapsed, yielding a tendency to increase the mean density. The cooling rate of the melt seemed not high enough to stabilize the cells as the polymer, having a poor heat conductivity, went through a rapid expansion process. The foam densities tended to increase with increases in the polymer concentration.

As might be expected, at comparable processing conditions, DDC foaming in microgravity would increase foam density through the decrease of cell sizes. And also, the rates of mass and heat transfer induced by decompression would change due to variations in flow mechanism.

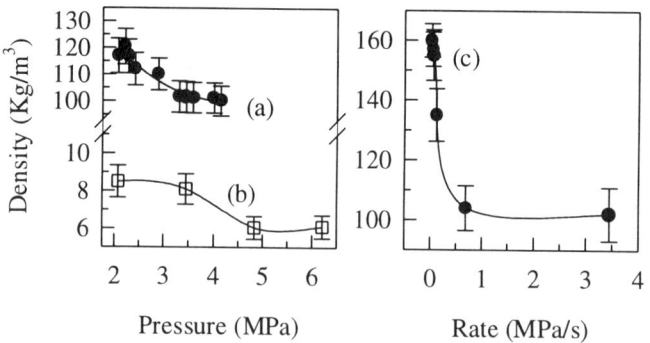

Figure 3. Mean densities (ρ_f) of the LDPE foams with MI=2.3 as a function of process conditions; (a) Mixed cell foams at $T_d = 110°C$ and $r_d = 2.3$ MPa/s, (b) Closed cell foams at $T_d = 140°C$, (c) Mixed cell foams at $T_d = 110°C$ and $P_d = 3.45$ MPa.

Molecular Structure and Crystallization

As-received LDPE resins displayed multiple melting peaks in DSC thermograms, with the highest peak around 113°C (Figure 4). The HDPE resin had an intense melting peak around 128°C. In the cooling operation, on the other hand, all the resins including the HDPE exhibited two distinct crystallization peaks. The lower peaks, occurred below 100°C, may be ascribed to crystallization of side-chain branches. Of all the materials, the LDPE with MI = 2.3 disclosed four melting peaks, implying that the resin may contain a second component or a high fraction of short and long chain branches of different molecular weights. In general, the long-chain branches have an influence on melt strength; as a result, at comparable molecular weights, LDPE reveals considerably higher melt strength than linear polyethylenes. The short-chain branches disrupt chain packing, lowering the melting temperature of crystallites formed by the main chains. It was seen that with DDC foaming, the melting peaks tended to diminish or move to high temperature regions, suggesting an increased crystal perfection in the formed polymers.

As-received LDPE resins exhibited crystallinities (X_c) around 45%, but the two LDPEs with lower MI had similar levels of X_c. This is presumably due to differences in branch point defects and topological constraints such as entanglements; both disrupt the crystallization process. During DDC foaming, the polymers appeared to undergo thermomechanical and solvent-induced crystallization that further increased their crystallinities. The DDC foams were

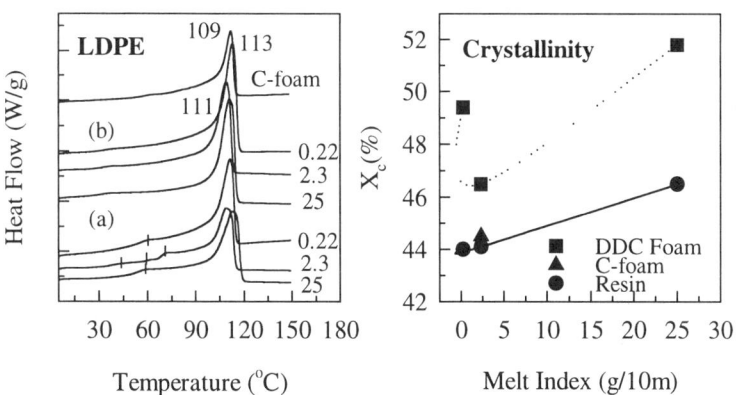

Figure 4. DCS thermograms and crystallinities of (a) as-received resins and (b) the DDC mixed cell foams, with a commercial foam (C-foam).

more crystalline than the resins, revealing X_c's around 50%. The foam with MI = 2.3, though, showed a minimum X_c, with a thermogram similar to that of a commercial LDPE foam; both revealed a rather broad melting endotherm around 109°C. The two polymers may have a similar molecular structure, which could influence resin foamability. The HDPE foams were more crystalline with X_c's around 60%.

It is reported that glasses crystallized in microgravity have a different microstructure with finer grains distributed more uniformly than equivalent samples crystallized on earth (20). Also, in liquid phase sintering of W-Ni-Fe metal alloy, the tungsten grains formed in microgravity are found to have smaller size, narrower dispersion, and lower population (21). It was speculated that the differences in crystalline structure were due to the absence of buoyancy driven convection, which was suspected to reduce the rates of crystal nucleation and growth, and to crystallization under stress-free conditions (20). In reduced gravity, we expect that DDC foaming would produce foamed polymers of lower crystallinity associated with the roles of surface tension and chemical homogenization, but the variations could be marginal.

Foam Morphology

The mixed cell foams that were produced at a decompression temperature (T_d) of 110°C generally revealed two different morphologies of granules and fibers, which reflect the mechanisms of their formation, fluctuations of the concentration as well as solvent- and flow-induced crystallization of the polymer (Figure 5). The inside closed cell walls changed their morphologies considerably with expansion. For some cells having diameters below 400 μm, the SEM images showed uniformly distributed granules on the wall surface, which had smooth surfaces and tended to align themselves along the circumferential direction (CD). During the foaming process, the polymer phase undergoes gas-liquid phase separation and rapid cooling, reducing the surface area or volume of foam cells. This tendency may lead to the formation of granulated surfaces inside the cell walls. On the wall strut, crystal lamellae with a thickness of a few microns were observed. Increases in expansion ratio stretched and flattened the surfaces, forming a fiber structure that resembled multilamellar habits and oriented apparently along the circumference of the cell membrane. A fiber network with thicknesses of ca. 10 μm was also seen. Around the fibers were dispersed granules with diameters of 1 to 10 μm that possessed either smooth or rough surfaces, characteristically similar to those isolated globules generated from dilute PE solutions through liquid-liquid phase separation (22). In contrast, the closed cell foams that were prepared at a T_d of 140°C exhibited only fiber structures in SEM micrographs.

179

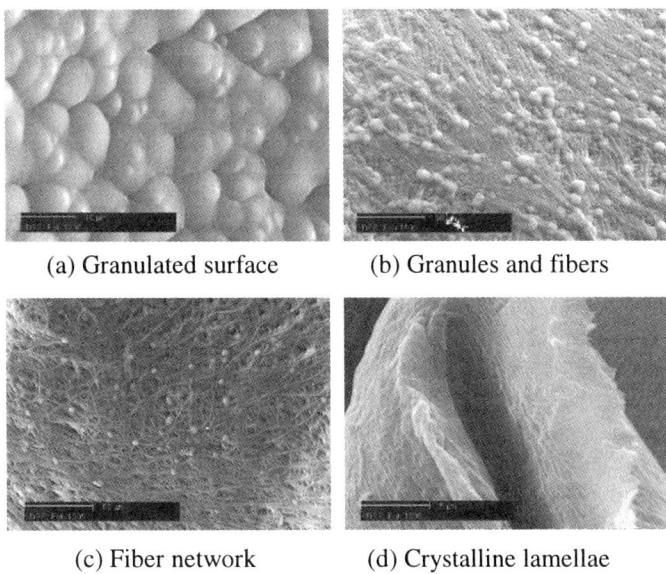

(a) Granulated surface (b) Granules and fibers

(c) Fiber network (d) Crystalline lamellae

Figure 5. SEM micrographs of the DDC mixed cell foams; (a)(b)(c): Closed cell walls, (d): Open cell walls.

The character of the mixed cell morphologies implies that the solutions may go through local fluctuations of the concentration, resulting in microdomains enriched in polymer, probably due to the nonuniformity of the initial state at the metastable regime. A subsequent depletion of the solvent phase may lead to the formation of granules on the wall surface. Fiber structures coexisting with the particles may, therefore, be generated. The smooth and rough surfaces of the particles, on the other hand, were attributed to crystallization of the polymer (22). In heterogeneous nucleation, the outwards growth of lamellae owing to foreign particles acting as heterogeneities was found to form the rough surface, while homogeneous nucleation resulted in a smooth, folded crystal surface. The lamellar thicknesses could also vary, possibly being much thinner in homogeneous nucleation.

The bubble cells eventually ruptured with further expansion, resulting in open cellular structures that exhibited diverse, quite well-developed lamellar habits with thicknesses of less than 1 μm, the texture of which was characterized by orientation of the chains along the flow direction of the volatile phase. This agrees in general with the previous results of poly(butylene terephthalate, PBT) foams that comprised open cell structures (8). Oriented lamellae interconnected laterally were also seen. In contrast, an LDPE specimen that crystallized from the melt under free of stress conditions was found to have a round-shaped lamellar-like structure on the surface with a mean diameter as small as ca. 20 Å.

Crystalline Structure and Orientation

WAXS 2θ scans and flat film photographs exhibit that LDPE closed cell foams contained distinct (110) and (200) crystalline peaks, both marginally strengthened with increasing expansion. However, the patterns changed little with MI. Figure 6 shows the pole figure of the (110) plane of LDPE closed cell foams, which is compared with those of the open cell foams of poly(butylene terephthalate, PBT) produced by DDC (8). In the PBT foams, the (100)α poles concentrated in the ND with spreads toward the ND-HD plane and the (010)α normals, which were not defined well, tended to disperse in bands in the FD-ND plane, indicating the presence of anisotropic texture. However, the LDPE (110) normals distributed quite uniformly in the planes of the foam.

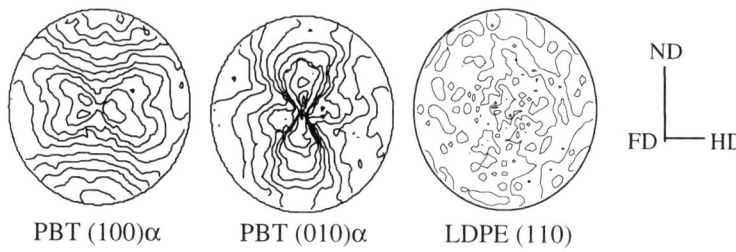

PBT (100)α   PBT (010)α   LDPE (110)

Figure 6. WAXS pole figures of DDC-produced open cell PBT and closed cell HDPE foams with different lattice planes.

Using Eq. (5), the orientation factors ($f_{i,j}^B$) have been computed and compared to those of a blown film inflated at a blow-up (BR) ratio of unity and a draw-down (DR) ratio of 20. As expected, the closed cell foams exhibited $f_{i,j}^B$ close to zero, suggesting that the overall texture resembles isotropy. This texture is responsible for the alignment of scattering units along the circumferential direction. In contrast, the blown film showed a preferential uniaxial orientation; the a- and c-axes are oriented primarily along the FD with the b-axis, revealing a very low orientation, close to the HD. The open cells of the PBT foams, on the other hand, comprised a weak biaxial chain orientation; the c-axis is oriented more along the FD and the a-axis, which is normal to the plane of the phenyl ring on the chain backbone, is tilted towards the equal biaxial line. It appears that with DDC foaming, the PBT chains in the open cell structures tend to orient in the FD with the phenyl rings aligning roughly parallel to the foam surface. The strong FD orientation is probably due to the flow character of the evaporating solvent phase in a container.

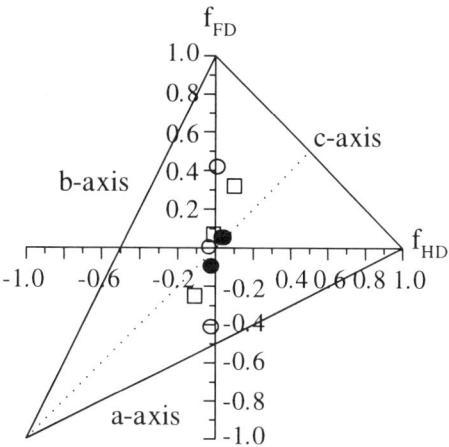

Figure 7. Crystalline orientation factors for DDC foams; Filled and open circles: HDPE closed cell foams and blown film respectively, Open square: PBT open cell foams.

Mechanical Property

The mechanical properties of semicrystalline polymer foams are related in a complex manner to the cellular morphologies of the foam and to the properties of the polymer matrix. Included in the former are foam density, cell structure and anisotropy, cell size and shape, and distributions of cell sizes, while the latter comprise molecular interactions, crystalline morphology and state of molecular orientation. At comparable levels of densities, foams with higher fractions and smaller sizes of closed cells are found to exhibit higher modulus and strength in tensile and compressive testing. Increasing density, on the other hand, strengthens foamed materials (23).

These previous findings are consistent with the results of the DDC foams (Figure 8). Foams tend to be stronger with increasing polymer concentration in solutions, probably due to the increased uniformity of cellular structure. The HDPE foams revealed higher tensile strengths because of their crystalline nature.

The DDC LDPE foams had appreciably higher strength-to-density ratios than a commercial LDPE foam. This mechanical behavior is clearly due to the structural character; the DDC foams revealed smaller cells that tended to be hierarchical along the thickness direction. Also, oriented crystals were found in the texture. The tensile drawing of oriented polymers tends to bend and stretch the covalent bonds of the molecules, resulting in a high Young's modulus and strength. In unoriented polymers, however, the deformation may increasingly involve the breakage of intermolecular forces, which imparts to the specimen an inferior modulus and strength.

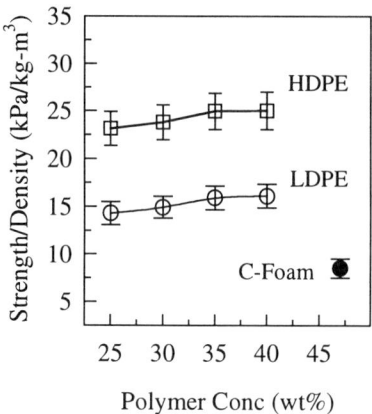

Figure 8. Tensile properties at maximum of DDC closed cell foams along the FD and of a commercial LDPE foam.

Conclusions

The DDC process has produced polymer foams of open and closed cell structures, hierarchical or uniform, with densities ranging from a low to medium level. The polymer phase contains the morphologies of granules, oriented fibers and network-like structures, which tend to strengthen foamed products. Lightweight, DDC-foamed materials, which possess enhanced strength-to-weight ratios, will be fabricated in a continuous manner. However, during foaming the solutions containing species that reveal a large difference in density tend to phase segregate, greatly narrowing the parameter space of the DDC process.

The microgravity environment can aid in providing idealized conditions for foaming. In particular, phase separation as well as layering and stratification effects are minimized so that uniform initial states can be achieved prior to decompression. Also, non-uniform heating produces convective effects that are absent in zero-g, allowing for the study of transient processes and resulting foams. These considerations are especially important when processing polymer blends and polymer/metal composites intended for high temperature applications. Therefore, DDC foaming in microgravity will serve not only the understanding of the underlying mechanisms of rapidly decompressed solid foams, but also may provide a new approach for producing structural materials to be used in space.

Acknowledgments

The authors gratefully acknowledge the NASA Office for Microgravity Materials Science for financial support (Grant No. NAG8-1461). The polymers investigated in this study were kindly offered by the Dow and Mobile Co.

References

1. Suh, K. W.; Webb, D. D. *Cellular Materials in Ency. Polym. Sci. Eng*; 2nd Eds, Wiley, NY, 1985
2. Munters, C. G.; Tandberg, J. G. U.S. Patent 2,023,204, 1935
3. McIntire, O.R. (to Dow) U.S. Patent 2,515,250, 1950
4. Rubens, L. C.; Griffin, J. D.; Urchick, D. (to Dow) U.S. Patent 3,067,147, 1962
5. Blades, H.; White, J. R. (to DuPont) U.S. Patent: 3,227,664, 1966
6. White, J. R.; Blades, H. (to DuPont) U.S. Patent: 3,542,715, 1970
7. Apfel, R. E.; U.S. Patent 5,384,203, 1995
8. Song, K.; Li, W.; Eckert, J. O.; Wu, D.; Apfel, R. E. *J. Mat. Sci.* **1999**, 34, 5387
9. Amon, M.; Denson, C. D. *Polym. Eng. Sci.* **1984**, 24, 1026
10. Arefmanesh, A.; Advani, S. G. *Polym. Eng. Sci.* **1995**, 35, 252
11. Geol, S. K.; Beckman, E. J. *AIChE J.* **1995**, 41, 357
12. Joshi, K.; Lee, J. G.; Shafi, M. A.; Flumerfelt, R. W. *J. Apppl. Polym. Sci.* **1998**, 67, 1353
13. Throne, J. L. *Thermoplastic Foams*; Sherwood Publishers; Hinckley, OH, 1996
14. Wunderlich B.; Cormier, C. M. *J. Polym. Sci. Phys.* **1967**, **5**, 987
15. White, J. L.; Spruiell, J. E. *Polym. Eng. Sci.*, **1980**, 20 247
16. Brandrup J.; Immergut, E. H. *Polymer Handbook* 3rd Eds., Wiley, NY, 1989
17. Harvey, E. N. *Photoplasmatologia*, **1954**, 2, 1

18. McManus, S. P., *Polymer Preprints*, **2000**, 41, 1056
19. Pojman, J. A., *NASA Microgravity Materials Science Conference*, **2000**, NASA/CP-1999-209092, 527
20. Day, D. E., Ray, C. S., *NASA Microgravity Materials Science Conference*, **2000**, NASA/CP-1999-209092, 153
21. Gokhale, A. M., Tewari, A., Mirabelli, T., *NASA Microgravity Materials Science Conference*, **2000**, NASA/CP-1999-209092, 271
22. Schaaf, P.; Lotz,B.; Wittman, J. C. *Polymer* **1987,** 28, 193
23. Meinecke, E. A.; Clark, R. C. *Mechanical Properties of Polymeric Foams*; Technomic, Westport, CT, 1973

Chapter 13

Prospects for the Study of Gas-Phase Polymerization and the Synthesis of Polymers Containing Nanoparticles in Microgravity

Yezdi B. Pithawalla, Junling Gao, and M. Samy El Shall*

Department of Chemistry, Virginia Commonwealth University, Richmond, VA 23284-2006

The study of gas phase and cluster polymerization is an important intellectual and technological frontier which promises unique results not only for a fundamental understanding of polymerization reactions, but also for the development of new materials with unique properties. In the past it has been almost impossible to study gas phase chain polymerization because the involatile product molecules condensed out of the gas phase. The application of microgravity to the study of gas phase polymerization is expected to result in a better control of the process and may also lead to important technological advances.

Despite the fact that solutions and bulk liquids are the preferred medium for many industrial and laboratory polymerization processes, our fundamental understanding of the polymerization reactions in solution remains limited. Under normal circumstances, polymerization is conducted in condensed systems [monomer liquid or solution] in which multiple reactions [initiation, propagation, chain transfer, termination, etc.] are occurring simultaneously *(1-3)*. Information regarding the exact nature of each mechanistic step and understanding of elementary events occurring in the course of polymerization remain largely unavailable. In solution the problem is further complicated by reaction within the solvent.

In order to arrive at a clearer understanding of these processes, the first step must be the elucidation of the detailed mechanism of the early stages of polymerization reactions in the gas phase, followed by studying these reactions within molecular clusters composed of few molecules [5-50], and finally within microdroplets of the monomer molecules. This approach provides an opportunity to study separate mechanistic steps such as initiation, propagation, inhibition and termination under controlled conditions. It also helps to bridge the gas phase and condensed phase chemistry and to provide unique environments for the synthesis of new polymeric materials with unusual properties.

Why Study Gas Phase Polymerization? & Why Microgravity?

In the gas phase it is possible to observe the direct formation, in real time, of product polymeric radicals, cations or anions of a chosen size. In contrast, in the liquid phase, one is usually forced to infer what has happened by the qualitative and quantitative analysis of the products. In the gas phase it is possible to suppress the process of recombinative or disproportionative termination, in the case of radical addition polymerization, as well as the process of chain transfer *(1-3)*. This makes it possible to study in isolation, the elementary steps [e.g. initiation, propagation, recombination, etc.] of the multistep polymerization reactions. It is also of interest to study polymerization in a poor solvent since the polymer is likely to adopt a globule rather than a coil configuration and the rate, and possibly the mechanism of reaction could be dramatically affected. However in a poor enough solvent, the solubility of the polymer is so small that quantitative measurement of the rate is difficult. Perhaps the poorest "solvent" is represented by the vapor so that the development of gas phase methods could contribute to studies of this kind.

From a practical point of view, gas phase polymerization can lead to the synthesis of defect-free, uniform thin polymeric films of controlled morphology and tailored compositions with excellent electrical and optical properties for many technological applications such as protective coatings and electrical insulators. For example, the polymeric species could be deposited from the gas phase in a size-selected manner on metal or semiconductor surfaces. Furthermore, gas phase polymerization can be easily coupled with the vapor phase synthesis of metal and semiconductor nanoparticles to generate novel hybrid materials, which combine several properties such as strength, elasticity, electrical conductivity and improved electro-optic performance. Of particular interest, is the design of composite materials consisting of polymer insulator surface layers and metallic or intermetallic cores. The properties of these materials can be tailored by altering the metal particles, their size and shape or their relative concentrations. Finally, gas phase polymerization eliminates the need for distillation, drying and solvent recovery and therefore, the operating costs and the environmental and atmospheric problems associated with these processes.

Given all the above advantages and promises of gas phase polymerization, it is unfortunate that the achievement of true homogeneous gas phase process for the synthesis of high molecular weight polymers is <u>very difficult</u>, and has not really been accomplished successfully in the past. The main problem is the extreme involatility of the large polymers, rendering it impossible to keep them in the gas phase. Several groups have recently focused on developing new methods to study polymerization reactions in the gas phase *(4-8)*, within isolated clusters in vacuum *(9-20)*, and within nucleating microdroplets in the vapor phase *(21-25)*. Although, these methods are ideal for the elucidation of the mechanisms and kinetics of the early stages of polymerization, they are limited as synthesis tools for large polymers because of the gravity effects. The application of microgravity to the study of gas phase polymerization and to the synthesis of advanced polymer / nanoparticles composites is expected to result in a better control of the process and may also lead to important technological advances.

In this chapter, we describe few examples of gas phase and intracluster oligomerization reactions for the study of the early stages of polymerization and present methods for the synthesis of polymeric materials containing ultrafine and nanoparticles. We also discuss several approaches and opportunities to study gas phase polymerization and synthesize novel polymeric materials in microgravity.

Experimental

A. Gas Phase Ion-Molecule Reactions

The systems under investigation include the radical cationic polymerization of olefin and diolefin monomers. In order to establish the mechanistic features of the initiation process, selective ionization of the initiator component is necessary to avoid direct ionization of the olefin monomer. For this reason, we use Resonant-Two-Photon-Ionization coupled with High Pressure Mass Spectrometry (R2PI-HPMS) where the mode of ionization is selective for the aromatic initiator component *(26,27)*. A schematic diagram of the experimental setup is shown in Figure 1. Gas mixtures are prepared in a 2 L glass flask heated > 100°C, and admitted to the ion source at selected pressures via an adjustable needle valve.

The laser beam is slightly focused within the center of the cell using a quartz spherical lens. The tunable UV laser light is generated by an XeCl excimer pumped dye laser [Lambda Physik LPX10 and FL-3002, respectively]. The reactant and product ions escape through a precision pinhole [200 μm] and are analyzed with a quadrupole mass filter. The same method can be used to study the reactions of monomer molecules with metal ions generated by laser vaporization / ionization of a metal target placed inside the high pressure cell as described in previous publications *(7,16)*.

B. Intracluster Reactions

For cluster polymerization studies, neutral clusters are produced using supersonic cluster beam techniques and the ionic polymerization reactions are initiated following the ionization of the molecular clusters either by electron impact or multiphoton ionization *(28)*. Other initiation mechanisms may involve metal-catalyzed reactions where metal ions and atoms are generated by laser vaporization and then allowed to react with the clusters of the selected moomer *(13,15,28)*. A schematic diagram of the experimental set up is shown in Figure 2. The essential elements of the apparatus are jet and beam chambers coupled to time-of-flight [TOF] and quadrupole mass spectrometers *(9,17,20)*. During operation 2%-10% of the monomer vapor in He or Ar carrier gas at a pressure of 2-3 atm is expanded through a conical nozzle [0.5 mm diameter] in pulses of 150-250 μs duration at repetition rates of 5-10 Hz. Following the ionization of the clusters, ion-molecule addition and elimination reactions take place within the cluster ions, resulting in a product ion distribution that reflects both the stability of the polymeric ions and the kinetics of the reaction *(28)*. The cluster ion beam is then analyzed using one of the mass spectrometers.

C. Polymerization Initiated by Plasma Laser Vaporization of Metals

A method for initiating cationic polymerization of bulk liquid monomers that leads to the incorporation of ultrafine metal particles into the polymer

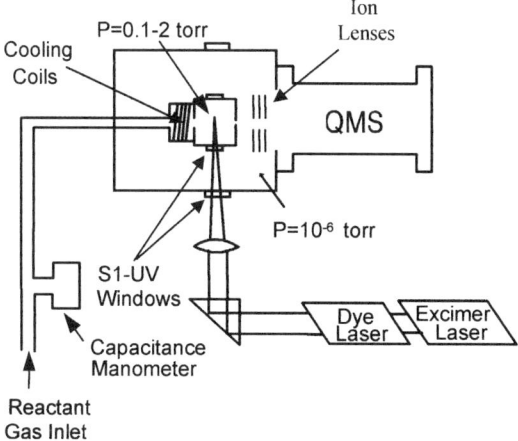

Figure 1. Experimental set up for the R2PI-HPMS system. "Reproduced with permission from reference 45. The ACS Division of Polymer Chemistry".

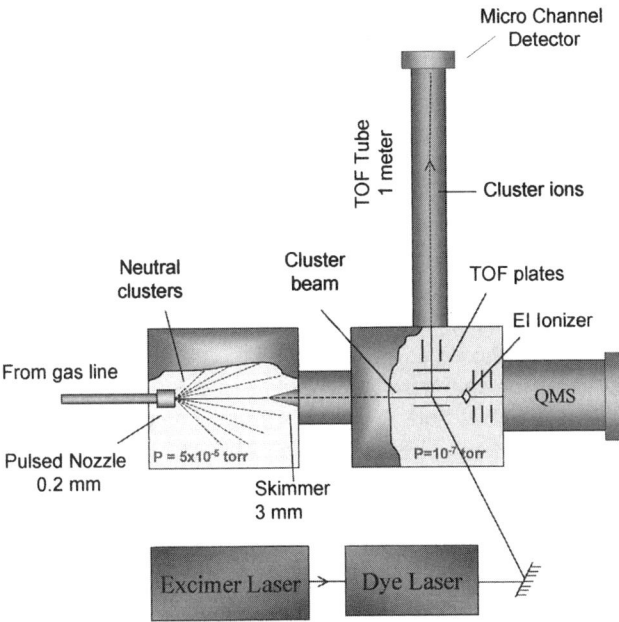

Figure 2. Cluster source with TOF and quadrupole mass spectrometers

matrices has been previously described *(29-32)*. Recently, we have described a technique to generate semiconductor, metallic and intermetallic nanoparticles from the vapor phase by a laser vaporization controlled condensation technique [LVCC] *(33-35)*. A schematic of the experimental set up is shown in Figure 3. The metal vapor is generated by pulsed laser vaporization using the second harmonic [532 nm] of Nd-YAG laser [15-30 mJ/pulse, 10^{-8} s pulse].

The metal target and the lower plate of the chamber are maintained at a temperature higher than that of the upper one [temperatures are controlled by circulating fluids]. The top plate can be cooled to less than 150 K by circulating liquid nitrogen. He is used as a carrier gas and the pressure can be varied from 10^{-3} to 10^3 torr. The nanoparticles are deposited on the cold top plate of the chamber. Plasma polymerization can be induced by using pure monomer vapor and applying appropriate electric field across the chamber plate. This method can generate metal nanoparticles incorporated in the polymeric materials with different morphologies.

Results and Discussion

A. Early Stages of the Radical Cation Polymerization of Isoprene and Propene

The ion chemistry of the olefins and diolefins plays an important role in the gas phase polymerization process *(7-9,28,36-38)*. These monomers can be induced to oligomerize and/or polymerize through bimolecular ion-molecule reactions in the gas phase *(28)*. The reactions can be initiated by an appropriate cation or radical cation which can transfer the charge to the selected monomer.

The Benzene/Isoprene Gas Phase System. In these experiments, the R2PI of benzene was obtained via the 6^1_0 transition at $\lambda=258.9$ nm. These photons create $C_6H_6^{+\bullet}$ ions with little excess energy [≤ 0.32 eV] and no benzene fragments were generated. Also, no direct photoionization of isoprene was observed in the absence of C_6H_6 in the reaction mixture. Figure 4 displays the R2PI mass spectrum obtained for a mixture of benzene and isoprene in N_2 carrier gas at a source pressure of 0.042 torr. Since the ionization potential, IP[isoprene] < IP[benzene], direct charge transfer from benzene$^{+\bullet}$ to isoprene takes place according to Reaction 1.

$$C_6H_6^{+\bullet} + C_5H_8 \rightarrow C_5H_8^{+\bullet} + C_6H_6 \qquad (1)$$

Following the generation of $C_5H_8^{+\bullet}$, it undergoes a series of well known addition and elimination reactions with neutral isoprene which produce the ion

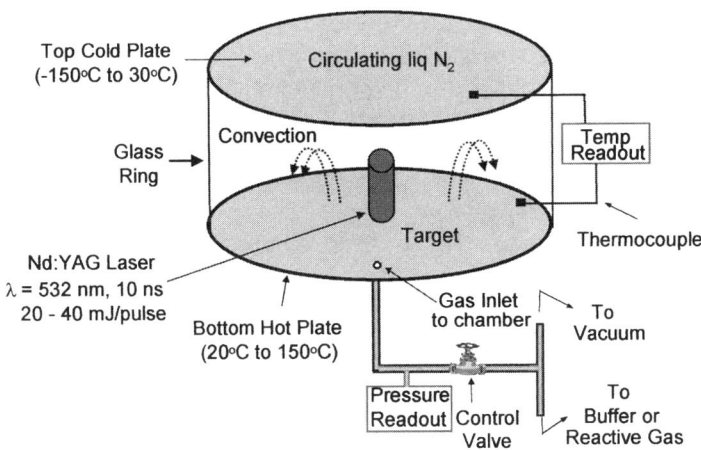

Figure 3. Experimental set up for the synthesis of nanoparticles coupled with plasma polymerization.

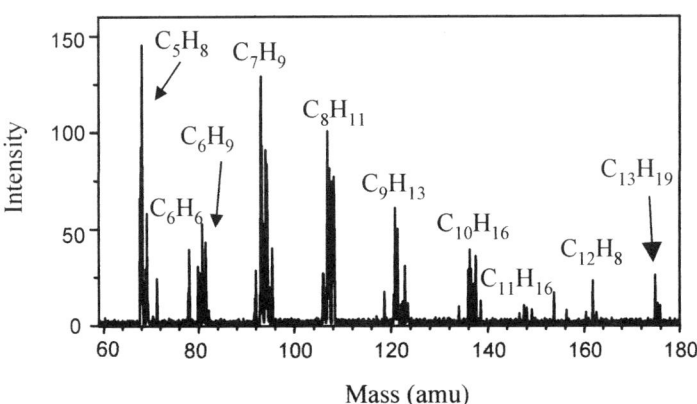

Figure 4. Mass spectrum obtained following the resonance ionization of benzene in a benzene / isoprene / N_2 mixture. "Reproduced with permission from reference 45. The ACS Division of Polymer Chemistry".

sequences C_xH_y with x =6-13 and y>x (9,37). The elimination products from the $C_5H_8^{+\bullet} + C_5H_8$ reaction involves the generation of the ions $C_9H_{13}^+$, $C_8H_{12}^{+\bullet}$, $C_8H_{11}^+$, $C_7H_{11}^+$, $C_7H_{10}^{+\bullet}$, $C_7H_9^+$, $C_7H_8^{+\bullet}$, $C_6H_9^+$, $C_6H_8^{+\bullet}$ [with m/z values of 121, 108, 107, 95, 94, 93, 92, 81 and 80; respectively]. These ions are produced through the elimination of the neutral species $CH_3^\bullet$, C_2H_4, $C_2H_5^\bullet$, C_3H_6, $C_3H_7^\bullet$, C_3H_8, $C_4H_7^\bullet$ and C_4H_8; respectively. The observed ions are consistent with the results obtained from electron impact ionization of gas phase clusters of isoprene *(9)*. They are also similar to those observed in bimolecular ion-molecule reactions of isoprene after a 300-ms reaction time in an ion trap *(37,38)*. Figure 5 represents the generation of reaction products in the system composed of benzene (B)/isoprene (I)/Ar at a source pressure of 0.58 torr at low and high concentrations of isoprene. The observed product channels include isoprene series $I_n^{+\bullet}$ with n =2,3; the adduct series $BI_m^{+\bullet}$ [m=1,2], and the elimination products mentioned above. The rapid decrease in the ion intensity of $C_5H_8^{+\bullet}$ due to the formation of the three product series is clearly observed at higher concentration of isoprene [Fig. 5-b]. It is well known that isoprene can undergo rapid Diels-Alder reactions with its molecular ion to form the dimer ion $C_{10}H_{16}^+$, which has the limonene structure *(9,37)*. This is supported by the observation of elimination products that are identical to the dissociation products of limonene *(9)*.

By increasing the concentration of isoprene in the reaction cell, ions corresponding to isoprene oligomers $(C_5H_8)_n^{+\bullet}$ up to n=4 could be observed. The significance of the generation of $C_5H_8^{+\bullet}$ is that it can initiate the radical cation polymerization of isoprene in absence of other fragment ions which are typically generated by electron impact and non-resonance multiphoton ionization.

The Benzene/Propene Gas Phase System. In this system, the aromatic initiator (C_6H_6) has an IP between the reactant's monomer (C_3H_6) and its covalent dimer (C_6H_{12}), i.e.; IP(C_3H_6) > IP(C_6H_6) > IP(C_6H_{12}). Therefore, direct charge transfer from $C_6H_6^{+\bullet}$ to C_3H_6 is not observed due to the large endothermicity of 0.48 eV and only the adduct $C_6H_6^{+\bullet}(C_3H_6)$ is formed. However, coupled reactions of charge transfer with covalent condensation involving the $C_6H_6^{+\bullet}$ ion and two C_3H_6 molecules are observed according to the overall process:

$$C_6H_6^{+\bullet} + 2\, C_3H_6 \rightarrow C_6H_{12}^{+\bullet} + C_6H_6 \qquad (2)$$

This reaction represents an initiation mechanism for the gas phase polymerization of propene since it results in the formation of the dimer radical cation ($C_6H_{12}^{+\bullet}$), which can sequentially add several propene molecules. The formation of this ion can make the overall process of charge transfer and covalent condensation significantly exothermic *(27)*. At higher concentrations of propene, the reaction products are the propene oligomers $(C_3H_6)^{+\bullet}{}_n$ with n = 2-7 and the adduct series $C_6H_6{}^{+\bullet}(C_3H_6)_m$ with m≤6.

Figure 6-a displays the time profiles corresponding to the sequential generation of the $(C_3H_6)^{+\bullet}{}_n$ series with n up to 5 (denoted as P_2, P_3, P_4 and P_5). The first event involves the decrease in the benzene ion signal (B) followed by the sequential generation of P_2, P_3, P_4 and P_5 series. Note that the propene ion $C_3H_6{}^{+\bullet}$ is not produced. Figure 6-b shows the normalized reaction profiles corresponding to the disappearance of the $C_6H_6{}^{+\bullet}$ ion signal (B), and the appearance of the two parallel channels $C_6H_6{}^{+\bullet}(C_3H_6)_m$ with m=1-3 (ΣBP_m), and the $(C_3H_6)^{+\bullet}{}_n$ series with n=2-6 (ΣP_n). The significance of the coupled charge transfer/covalent condensation reactions is that the overall process leads exclusively to the formation of condensation products $(C_3H_6)^{+\bullet}{}_n$ and avoids other competitive channels in the ion-molecule reactions of propene. For example, the reactions of $C_3H_6{}^{+\bullet}$ with neutral C_3H_6 involve several channels starting with the formation of the $C_3H_7{}^+$, $C_4H_7{}^+$ and $C_4H_8{}^{+\bullet}$ ions and their association products.

The Benzene/Isoprene/Propene Gas Phase System. In this ternary system, following the selective ionization of benzene direct and coupled charge transfer reactions can take place according to reactions 1 and 2, respectively. The observed reaction products include $IP_n{}^{+\bullet}$ (n≤4), $I_2P_n{}^{+\bullet}$ (n≤2), $P_n{}^{+\bullet}$ (n=2-5), $BP_n{}^{+\bullet}$ (n≤2) and the elimination products from bimolecular $C_5H_8{}^{+\bullet}$ + C_5H_8 reactions. The first two series are new product channels and represent the copolymerization involving isoprene radical cations and propene molecules. The other three products have been observed in the benzene$^{+\bullet}$/propene system and the benzene$^{+\bullet}$/isoprene systems, respectively.

The effect of source pressure on the product channels in the ternary system is shown in Figure 7. At higher source pressure, the product channels $IP_n{}^{+\bullet}$, $P_n{}^{+\bullet}$, and $I_2P_n{}^{+\bullet}$, along with elimination products of isoprene are dominant. At higher concentrations of propene, the formation of $P_n{}^{+\bullet}$ becomes very significant and higher order ions (n=2-6) can be observed.

As a reaction initiating species, isoprene radical cation $C_5H_8{}^{+\bullet}$ can initiate co-oligomerization by direct attack on a neutral propene monomer which may result in a cyclic product as shown below.

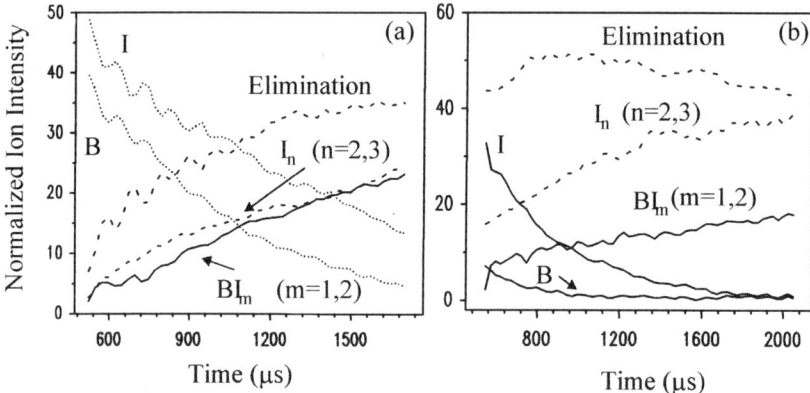

Figure 5. Time profiles of reactant and product ions obtained at a source pressure of 0.55 torr (Ar) and concentration number density of Isoprene of (a) 4.5×10^{11} cm^{-3} and (b) 3.8×10^{12} cm^{-3}

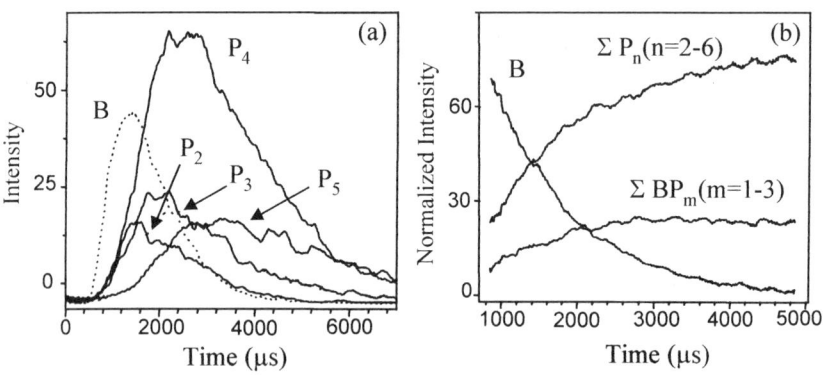

Figure 6 (a) Raw Ion time profiles and (b) Normalized ion profiles due to reactant and products in the benzene$^{+\bullet}$/propene system. "Reproduced with permission from reference 45. The ACS Division of Polymer Chemistry"

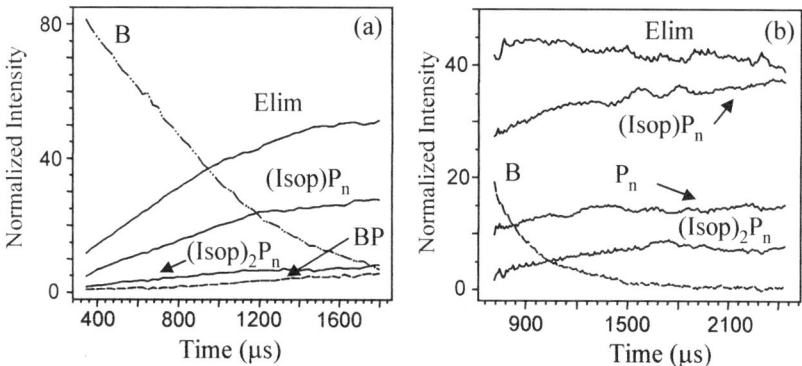

Figure 7. Normalized ion time profiles in the ternary benzene$^{+\bullet}$/ propene/ isoprene system at a source pressure of (a) 0.25 torr and (b) 0.8 torr (Ar).

This isoprene-propene adduct ($C_8H_{14}^{+\bullet}$) can undergo a further reaction with a second C_3H_6 molecule to form $IP_2^{+\bullet}$, thus leading to co- oligomerization by the sequential additions of C_3H_6 molecules. On the other hand, the structure of the $I_2P_2^{+\bullet}$ adduct appears to be unreactive as indicated by the absence of ions $I_2P_n^{+\bullet}$ with n ≥3 at high concentrations of propene. The stability of the $I_2P_2^{+\bullet}$ ion could be a possible reason for the termination of further additions of propene.

Intracluster Reactions: Benzene/Isoprene and Benzene/Isoprene/Propene Clusters. The mass spectra of the binary benzene-isoprene and the ternary benzene-isoprene-propene clusters following the multiphoton ionization at 248 nm are shown in Figures 8-a and 8-b, respectively.

The results demonstrate that intracluster ion-molecule reactions can take place following the ionization of benzene in the clusters. The resulting ions are similar to gas phase products observed in the high pressure system. For example, the major series of ions in the benzene (B)-isoprene (I) clusters is $I_2^+B_m$, which is consistent with the rearrangement of the isoprene dimer ion $I_2^{+\bullet}$ to the stable limonene structure. In the ternary system, the series $I_n^{+\bullet}$, $IP_n^{+\bullet}$, $P_n^{+\bullet}$, and $I_2P_n^{+\bullet}$ are observed with significant intensity consistent with the gas phase results.

B. Synthesis of Polymeric Materials by Laser Vaporization - Induced Polymerization

Current methods for the preparation of polymer / metal composites are generally focused on plasma polymer thin films and include simultaneous plasma etching and plasma polymerization *(41,42)*, simultaneous evaporation of polymer and metal from separate sources or simultaneous plasma polymerization and metal evaporation from high temperature crucibles *(43)*. The combination of laser vaporization / ionization of metals with the very fast propagation rates characteristic of ionic polymerization offers great promise and advantages in this regard. The potential advantages of laser vaporization include solvent free environment, variable metal vapor flux depending on the laser power and sequential or simultaneous evaporation of several metals.

As briefly mentioned in the experimental section, we have discovered a novel technique for polymerization using metal ions generated in the gas phase by laser vaporization techniques *(29-32)*. The ions are pulled toward the monomer liquid by applying appropriate electric fields across the reaction chamber. Using this method, high molecular weight polymers [10^6 units, polyisobutene] have been synthesized *(29,30)*. The polymeric materials also contain micron- and submicron-sized metal particles as shown in Figure 9-a. The surface morphology of the polymer films obtained is dependent on experimental conditions such as laser power, temperature, pressure and electric

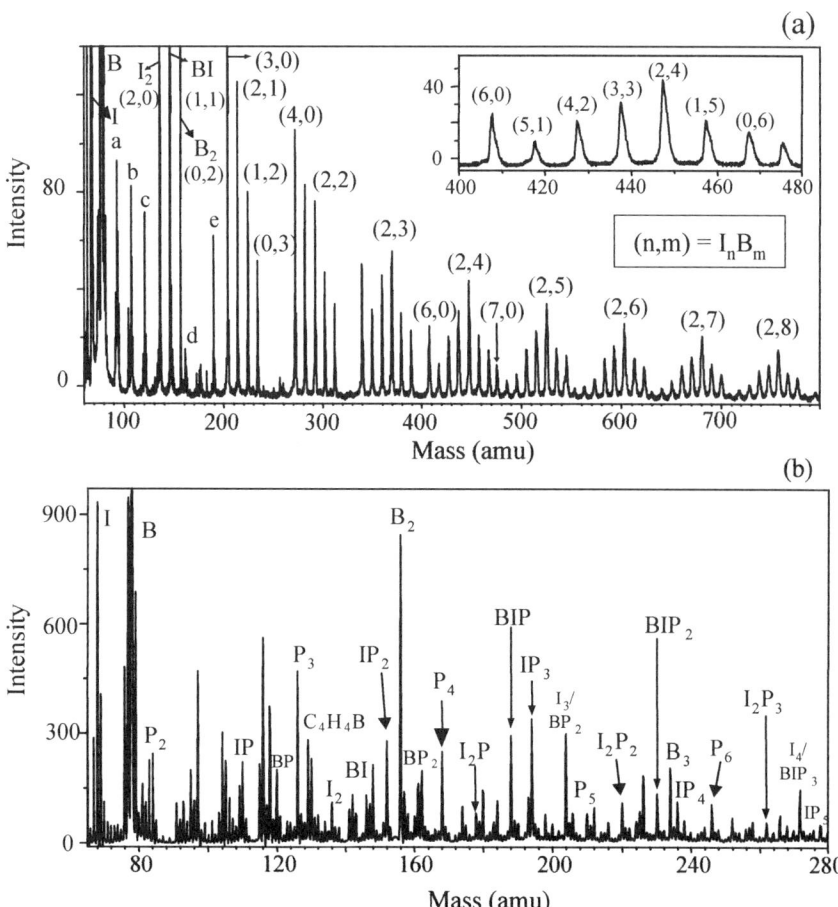

Figure 8. TOF mass spectra of (a) benzene-isoprene (I_nB_m) clusters. Peaks with m/z 93 (C_7H_9), 107 (C_8H_{11}), 121 (C_9H_{13}), 161 ($C_{12}H_{17}$), 189 ($C_{14}H_{21}$) are denoted as a,b,c,d and e, respectively. (b) benzene(B)-isoprene(I)-propene(P) clusters.

Figure 9. SEM of (a) polyisobutene containing ultrafine Cu particles and (b) polyisobutene microbeads containing Ni nanoparticles.

field strength. In another application, laser vaporization of a Ni target in presence of a high pressure of pure isobutene vapor [10^3 torr] produces a mixture of Ni nanoparticles and some polymeric microbeads as shown in Figure 9-b. We believe that the control of these microstructures is hampered by the gravity effects which introduce convection and nonhomogeneous conditions. We propose to carryout these experiments in space, and we expect to be able to control the microstructures and the size distribution of the polymer beads or the polymer containing metal nanoparticles. In the flight experiments, we will explore the use of plasma polymerization of the thin monomer liquid films as a simple alternative to laser vaporization. These new directions which are explored for the first time, could lead to the production of new polymeric films with unique material properties.

Implications for Gas Phase Polymerization in Microgravity.

Gas phase polymerization can be initiated by photon irradiation [using a high-pressure Hg / Xe UV lamp] or by electron beam ionization [using a heated filament] of the monomer vapor as we demonstrated in the ground-based experiments. Plasma polymerization can also be used by applying RF-field between two electrodes within the monomer vapor. The size of the product polymer particles is limited in normal gravity by gravitational settling. Polymerization can be terminated by recombination, or by the injection of inhibitors. The growth of the particles will be monitored by scattering of a test laser beam introduced perpendicular to the photoinitiating UV light or by a CCD camera. The product particles will be collected and the particle size distribution will be analyzed microscopically.

Polymerization within small-compartmentalized volumes of monomer liquid droplets is widely used in suspension and emulsion polymerization (44). An analogous new process of compartmentalized polymerization in liquid droplets suspended in the vapor can be used in microgravity since large nucleating droplets can be suspended indefinitely, without the presence of the solvent. In these experiments, we will generate an aerosol of monomer droplets by mechanical dispersion from a reservoir. We will then use pulsed irradiation by a UV lamp to initiate polymerization in the monomer vapor and/or within the monomer droplets. Initiation in the aerosol droplets will be achieved by two mechanisms. First, the UV irradiation will generate radicals in the gas phase that will proceed to polymerize. The living polymerizing particles can then be absorbed into a monomer droplet and continue polymerizing. Alternatively, the UV irradiation can generate radicals within already formed droplets.

In the polymerization experiments, pulsed additions of solvent or inhibitor vapor will be employed in order to study the solvent and inhibitor effects. The addition of inhibitors will also be used to stop the polymerization reaction after variable delay time from the initiation by UV pulses. This will stop the gas phase or intra-droplet polymerization before the products enter the liquid pools. The time-resolved progress of the polymerization process will be monitored in situ by laser light scattering.

Under microgravity conditions it would be possible to keep the growing polymers in the supersaturated vapor where they continue to propagate to larger sizes and thus kinetic information on the growing large polymers can be obtained. This is also a method to produce very narrow size distributions of polymers since few chains are growing in the vapor phase thus chain transfer and termination reactions can be minimized.

Acknowledgment. The authors gratefully acknowledge financial support from the National Science Foundation [Grant CHE 9816536] and the NASA Microgravity Materials Science Program [Grant NAG8-1484].

Literature Cited

1. *Introduction to Polymers;* Young, R. J.; Ed.; Chapman and Hall: New York, 1986.
2. *Principles of Polymerization;* Odian, G.; Ed.; 2nd Edition; John Wiley & Sons: New York, 1986.
3. *Carbocationic Polymerization;* Kennedy, J. P.; Marechal, E.; Eds.; John Wiley & Sons: New York, 1982.
4. Wang, J.; Javahery, G.; Petrie, S.; Bohme, D. K. *J. Am. Chem. Soc.* **1992**, 114, 9665.
5. Guo, B. C.; Castleman, A. W., Jr. *J. Am. Chem. Soc.* **1992**, 114, 6152.
6. Brodbelt, J. S.; Liou, C. C.; Maleknia, S.; Lin, T. J.; Lagow, R. J. *J. Am. Chem. Soc.* **1993**, 115, 11069.
7. Daly, G. M.; El-Shall, M. S., *J. Phys. Chem.* **1994**, 98, 696.
8. Meot-Ner (Mautner), M.; Sieck, L. W.; El-Shall, M. S.; Daly, G. M. *J. Am. Chem. Soc.* **1995**, 117, 7737.
9. El-Shall, M. S.; Marks, C. *J. Phys. Chem.* **1991**, 95, 4932.
10. El-Shall, M. S.; Schriver, K.E. *J. Chem. Phys.* **1991,** 95, 3001.
11. Coolbaugh, M. T.; Whitney, S. G.; Vaidyanathan, G.; Garvey, J. F. *J. Phys. Chem.* **1992**, 96, 9139.
12. Tsukuda, T.; Kondow, T. *J. Phys. Chem.* **1992**, 96, 5671.
13. Daly, G. M.; El-Shall, M. S. *Z. Phys. D.* **1993**, 26S, 186.

14. Tsukuda, T.; Kondow, T. *J. Am. Chem. Soc.* **1994**, 116, 9555.
15. Daly, G. M.; El-Shall, M. S. *J. Phys. Chem.* **1995**, 99, 5283.
16. Daly, G. M.; Pithawalla, Y. B.; Yu, Z.; El-Shall, M. S. *Chem. Phys. Lett.* **1995**, 237, 97.
17. El-Shall, M. S.; Daly, G. M.; Yu. Z.; Meot-Ner (Mautner), M. *J. Am. Chem. Soc.* **1995**, 117, 7744.
18. Desai, S. R.; Feigerle, C. S.; Miller, J. *J. Phys. Chem.* **1995**, 99, 1786.
19. Pithawalla, Y. B.; Gao, J.; Yu, Z.; El-Shall, M. S. *Macromolecules.* **1996**, 29, 8558.
20. El-Shall, M. S.; Yu, Z. *J. Am. Chem. Soc.* **1996**, 118, 13058.
21. Reiss, H. *Science.* **1987**, 238, 1368.
22. El-Shall, M. S.; Bahta, A.; Rabeony, H. M.; Reiss, H. *J. Chem. Phys.* **1987**, 87, 1329.
23. El-Shall, M. S.; Reiss, H. *J. Phys. Chem.* **1988**, 92, 1021.
24. Schmidtt, J. L. *J. Chem. Phys.* **1988**, 89, 5307.
25. El-Shall, M. S.; Rabeony, H. M.; Reiss, H. *J. Chem. Phys.* **1989**, 91, 7925.
26. Daly, G. M.; Meot-Ner (Mautner), M.; Pithawalla, Y. B.; El-Shall, M. S. *J. Chem. Phys.* **1996**, 104, 7965.
27. Meot-Ner (Mautner), M.; Pithawalla, Y. B.; Gao, J.; El-Shall, M. S. *J. Am. Chem. Soc.* **1997**, 119, 8332.
28. Pithawalla, Y. B.; El-Shall, M. S. In *"Solvent-Free Polymerizations and Processes: Minimization of Conventional Organic Solvents"*, Long, T. E.; Hunt, M. O., Eds.; ACS Symposium series 713; 1998, pp 232-247.
29. Vann, W.; El-Shall, M. S. *J. Am. Chem. Soc.* **1993**, 115, 4385.
30. Vann, W.; Daly, G. M.; El-Shall, M. S. In *"Laser Ablation in Materials Processing: Fundamentals and Applications"*, Braren, B.; Dubowski, J.; Norton, D., Eds.; Materials Research Society Symposium Proceedings Series, 1993, 285, 593.
31. El-Shall, M. S.; Slack, W. *Macromolecules,* **1995**, 28, 8546.
32. El-Shall, M. S. *Applied Surface Science,* **1996**, 106, 347.
33. Li, S.; Silvers, S. J.; El-Shall, M. S. *J. Phys. Chem B.* **1997**, 101, 1794.
34. El-Shall, M. S.; Li, S. In *"Advances in Metal and Semiconductor Clusters"*, Duncan, M. A. Eds.; Jai Press: London, 1998; 115-177.
35. Li, S.; Germanenko, I. N.; El-Shall, M. S. *J. Cluster Science.* **1999**, 10, 533.
36. Groenewold, G. S.; Gross, M. L. *J. Am. Chem. Soc.* **1984**, 106, 6569.
37. Kascheres, C.; Cooks, R. G. *Anal. Chim. Acta.* **1988,** 215, 223.
38. Vincenti, M.; Horning, S. R.; Cooks, R. G. *Org. Mass Spectrom.* **1988**, 23, 585.
39. Bowers, M. T.; Elleman, D. D.; O'Malley. R. M.; Jennings, K. R. *J. Phys. Chem,* **1970,** 74, 2583.
40. Bowers, M. T.; Aue, D. H.; Elleman, D. D. *J. Am. Chem. Soc.* **1972**, 94, 4255.

41. Perrin, J., Despax, B., Hanchett, V. and Kay, E. *J. Vac. Sci. Technol. A* **1986**, 4, 46.
42. Kay, E. *Z. Phys. D.* **1986**, 3, 251.
43. Heilmann, A.. and Hamann, C. in "*Electronic Properties of Polymers*," (eds. Kuzmany, H., Mehring, M., Roth, S.) 429-433 (Springer-Verlag, Berlin Heidelberg, **1992**).
44. Hansen, E. K.; Ugelstad, J. in "*Emulsion Polymerization*", Piirma, I. Ed., Academic Press, New York, 1982, pp. 51-92.
45. Pithawalla, Y. B.; Gao, J. and El-Shall, M. S. *Polmer Preprints* **2000**, 41, 1074.

Chapter 14

Production of Bacterial Polyesters in Simulated Microgravity

Radhika Thiruvenkatam[1] and Carmen Scholz[2],*

Departments of [1]Chemical Engineering and [2]Chemistry, University of Alabama at Huntsville, John Wright Drive, Huntsville, AL 35899

The synthesis of bacterial polyesters was demonstrated in simulated microgravity, by growing *Azotobacter vinelandii UWD* in the NASA-Bioreactor. Bacterial growth in simulated microgravity differed significantly from that observed in conventional shake flask experiments. Cells tended to grow in a cluster-like pattern. Bacterial cells started to produce polymer immediately after exposing them to conditions of simulated microgravity, no lag time was observed. Within the first 24 hours of the fermentation, bacterial polyester production was threefold higher in the bioreactor than in conventional shake flasks used as control. In an effort to differentiate between the effects of microgravity and the diffusion based aeration of the bioreactor, a gas supply profile was developed that led to similar amounts of dissolved oxygen in the bioreactor and in shake flask experiments. The bacterial growth behavior in the bioreactor was studied by monitoring the glucose and oxygen consumption and was compared to that in conventional shake flask experiments.

Introduction

Our society is challenged with the need for an increasing number of consumer goods and with the need for their disposal after their usability has been exhausted. New resources for materials production have to be explored and it is imperative to establish alternative disposal and recycling procedures. The aspects of degradability and ability to recycle become even more crucial when a closed system, like a spaceship leaving the Earth or a habitat on a distant planetary body, is considered. Materials production and management of solid waste become an imminent and intertwined problem when the production of materials, their use and eventual disposal are confined to such a limited space. Due to stringent restrictions in mass that a mission can carry from the Earth, materials must be chosen judiciously and, where possible, should be designed for several recycling cycles. The more diversity, with respect to the materials properties, that each use-recycling step allows, the greater the self-sufficiency of the travelling system.

In the near future, space exploration will continue to focus on Lower Earth Orbit, i.e. on board the Space Shuttle and the International Space Station. The latter one is a unique platform to test the reliability of a closed travelling system with no immediate and only limited access to supplies from Earth. The success of the International Space Station will, in part, depend on the reliability of the Advanced Life Support System operating in the station. Once a reliable system has been established that guarantees the safety and provision of the crew with food, and manages the recycling of carbon dioxide and waste that is generated in the course of human activity in space, missions will start to focus on deep space exploration. NASA plans human missions to planetary bodies in our solar system, with the Moon and Mars as immediate destinations after 2010. Even though, missions lasting longer than 600 days, the minimum for a round trip to Mars, are beyond today's technical capabilities, it is imperative to explore now avenues that will result in the realization of these daring plans. Some of the most important issues are the protection of the crew from cosmic radiation, the regeneration of air, water and food, and a management of solid waste, that guarantees optimum resource recovery and a safe environment in the vehicle or habitat. It will be mandatory that medium and long-term space missions produce their food on board, crops have been already identified that will be suited for vehicular activities, the near future goal, and planetary activities in the more distant future.

One way to combine the management of generated waste and the recovery of resources from solid waste, is the utilization of biological systems. They are unique for this task since they are characterized by a set of properties, which distinguish them from any other "production" facility. They regenerate, are energy and size efficient and they can adapt to changing environments. In addition, careful interference with biological pathways results in the formation of tailored products designed for specific properties and uses. Special emphasis is placed upon microorganisms, which are extremely versatile with respect to their carbon sources and adaptability to changing environments. Microorganisms are known to degrade organic material ranging from cellulose, to crude oil and toxic waste. Additionally, with the tools of recombinant DNA techniques in place, the capabilities of microorganisms can be expanded beyond their natural capacity and tailored to almost any task. Microorganisms not only degrade organic material, they are also capable of producing useful materials, and they are already exploited for this purpose; e.g. yeasts secrete biologically active glycolipids in comparatively large quantities (*1,2*), other microorganisms produce surfactants that are useful in the degreasing of machinery (*3,4*). Bacteria, in particular *E. coli*, have been studied extensively and the application of recombinant DNA techniques led to their exploitation for the synthesis of biomedically relevant proteins (*5,6*). Other bacterial strains, as elaborated below, produce structural materials, biopolyesters, in significant amounts. Based on their versatility, ability to regenerate, ease of handling and most importantly the fact that they thrive in an aqueous environment, thus, omitting the need for harsh chemicals or extreme synthesis conditions, microorganisms can be an ideal candidate for waste management and materials production onboard a spaceship. Ideally, waste will be transformed into a new and value-added material that can undergo several steps of recycling once its usability is exhausted.

As indicated above, microorganisms could be useful for several different types of application onboard a spaceship. The present study demonstrated the feasibility of producing biopolyester by microorganisms in simulated microgravity. The biological system under investigation in this study was *Azotobacter vinelandii UWD*, a bacterial strain known to produce poly(β-hydroxybutyrate), PHB, a biopolyester of the family of poly(β-hydroxyalkanoates), PHAs. Microorganisms have been studied in microgravity before, in particular on the Spacelab missions. All of these studies focused however, on microorganisms associated with the human body or pathogens. Most of these studies found a different growth behavior in microgravity as compared to bacterial growth on Earth. Space flight experiments have shown that fermentation experiments performed in microgravity were characterized by shortened lag times and increased final cell densities (*7-13*). The mechanisms leading to enhanced proliferation are not fully understood yet. It is assumed that the quiescent conditions in microgravity might cause fewer disruptions of the cell-cell interactions leading to genetic exchange. Even though the mechanism is not yet understood, the influence of increased radiation in low orbit can be ruled out as the determining factor. Fermentation experiments that were performed in the Shuttle using a reference centrifuge to simulate 1g gravity yielded a bacterial growth behavior that was comparable to the growth behavior observed on Earth (*14-17*).

Biopolyesters produced by bacteria

A wide variety of microorganisms produces PHAs for the purpose of carbon and energy storage (*18,19*). The most common biopolyester and the one produced by the majority of biopolyester-producing microorganisms is PHB, see Figure 1. PHB is produced by aerobic as well as anaerobic microorganisms. Polymerase enzymes become activated by environmental changes such as the shortage of one or more nutrients. Under experimental conditions, nitrogen and/or oxygen limitation are employed to initiate polymerase activity. Bacterial polyesters are unique with respect to some of their properties, they are isotactic, biodegradable and biocompatible. Moreover, they are highly UV-resistant and thermoplastic. All chiral Carbon atoms in the polymer backbone are in perfect R-configuration, see Figure 1.

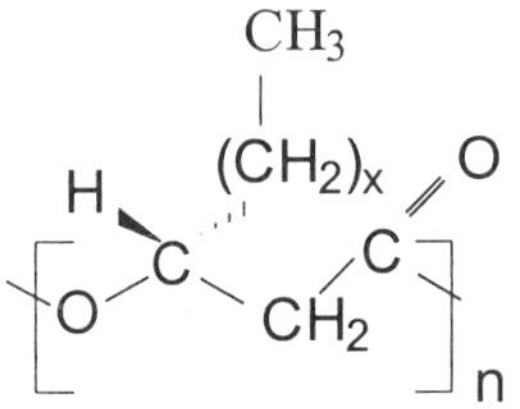

$x = 0, 1$: short chain PHA
e.g. $x = 0$: Poly(β-hydroxubutyrate) (PHB)

$x > 2$: long-chain PHA

Figure 1: Structure of bacterial polyesters

Perfectly isotactic polymers are not readily synthesizable by conventional synthetic routes; this is a characteristic that can be only achieved by enzyme catalysis. The isotacticity of bacterial polyesters is attributed to the enzymatic synthesis at the active site of the polymerase enzyme. Among others, isotacticity is essential for biodegradability (*20*), and biocompatibility (*21*). The isotacticity is also largely responsible for the physical properties of the material. The melting point of the polymer is about 170 °C. Due to its high crystallinity (> 80%), the polymer crystallizes readily in helix conformation, and the presence of large spherulitic crystals that form upon cooling or solvent evaporation, makes PHB a rather brittle and stiff material. The incorporation of hydroxyvalerate repeat units into PHB leads to the formation of a biocopolyester: poly(β-hydroxybutyrate-co-valerate) (PHBV) with properties reminiscent of polypropylene. The incorporation of HV units influences the materials properties, e.g. melting point and the enthalpy of fusion go through a minimum at 35 mole% HV units. PHAs can be easily extracted and used to fabricate useful materials and objects. PHB and PHBV have been tested for a variety of applications, some of which are listed in Table 1.

Table 1: Commercial application of PHB and PHBV

Bio-polyester	Application	US Patent/Reference
PHB and PHBV	Packaging materials (bottles, containers)	Baptist, 3,072,538 (1963) Baptist, 3,107,172 (1963) Webb, A. 4,900,299 (1990)
	Films, laminates Personal hygiene articles	Holmes, P.A. 4,620,999 (1986) Noda, I. 5,536,564 (1996) Noda, I. 5,489,470 (1996) Shiotani, T. 5,292,860 (1994)
	Fibers for nonwoven fabrics Hot-melt adhesive Biomedical applications	Steel, M.L. 4,603,070 (1986) Kauffman, T. 5,169,889 (1992) Scholz, C. (22)

The current draw-back to the use of PHAs on Earth is their cost of production. Even though it is an environmentally benign process, currently it is cheaper to produce thermoplastic materials with similar mechanical properties from petroleum. On a long-term space mission however, there is no access to raw materials from Earth, and in addition, an accumulation of organic waste occurs. These waste products can not be discarded since it is imperative to conserve and transform the energy inherent in the waste. Moreover, any process performed onboard a spacecraft needs to comply with safety regulations and should be as non-intrusive as possible. Since fermentation processes are performed at ambient temperatures and do not require the use of harsh chemicals, they are an ideal approach to the production of the structural and other materials, potentially from waste products.

Simulation of microgravity using the NASA- Bioreactor

Data and information that are available today on the bacterial production of biopolyesters were gathered so far on Earth, i.e.1g gravity. To test the bacterial capability of producing biopolyesters on a space station or onboard a traveling spacecraft, that is, in microgravity, it was necessary to mimic some aspects of microgravity in ground-based experiments. The NASA-Bioreactor, see Figure 2, was originally designed for the investigation of unrestricted, three-dimensional mammalian cell development and for the protection of mammalian cells from extensive shear forces during take-off and landing of the Space Shuttle. The cylindrical device, 95 mm in diameter and 8 mm in height, is engineered to accommodate the unrestricted development of a three-dimensional cytoarchitecture. Cells and cell aggregates rotate at the same speed as the vessel itself, thus constant randomization of the gravity vector subjects the cells to a continuous state of simulated free fall. The rotation of the bioreactor prevents settling of the cells, and at the same time maintains a quiescent environment that can not be accomplished by conventional stirring or experiments in shake flasks. Damaging turbulence is minimized in the bioreactor, and cells do not collide with vessel walls or any other damaging objects, such as stirrer blades. Destructive shear forces are minimized because this system has no impellers, bubbles, or agitators. All transport processes are diffusion based. The system is aerated with moistened air, oxygen, or a mixture thereof through membrane guaranteeing bubble free gas supply.

Figure 2: Cell mass accumulating in the NASA Bioreactor

Experimental

Strain information: The bulk of this study was performed using the bacterial strain *Azotobacter vinelandii UWD* (ATCC 53799). Stock cultures of *A. vinelandii UWD* were maintained at a growth medium (ATCC medium 1771) supplemented with 20 % (v/v) glycerol at –80°C. For polymer production cells were grown under nitrogen limiting conditions. One mL of the stock culture was aseptically inoculated into the minimum medium composed of K_2HPO_4 0.2g/L, KH_2PO_4 0.8g/L, $CaSO_4$ x $2H_2O$ 0.1g/L, $MgSO_4$ x $7H_2O$ 0.2g/L, CH_3COONH_4 1.1g/L, $FeSO_4$ x 7 H_2O 0.005g/L, Na_2MoO_4 x $2H_2O$ 0.00025g/L, Ferric Ammonium Citrate 0.06g/L, Yeast Extract 0.5 % wt/vol, Glucose 3% wt/vol. *A.vinelandii* was grown in the NASA-Bioreactor and for control experiments in conventional shake flask experiments. Typically, 250 mL medium was prepared, medium and glucose were autoclaved separately and combined after they cooled to room temperature. After the medium was inoculated with 1 mL of a rapidly thawed stock culture, 55mL was transferred aseptically into the bioreactor through the fill port. The bioreactor was tilted during the transfer and filled completely to obtain an air bubble-free medium. The remaining culture served as control experiments and was grown in a dented 500 mL Erlenmeyer flask. The bioreactor was placed in an incubator at 30°C and rotated at speeds increasing from 8 to 31 rpm as the cell mass grew. The bioreactor was aerated with hydrated air or hydrated oxygen. It was imperative to moisten the gas prior to introduction to prevent the evaporation of the medium that would lead to the formation of disturbing air-bubbles in the bioreactor. Shake flask experiments were performed in a New Brunswick Shaker Incubator at 200 rpm and 30°C.

Analysis: Bacterial cell growth in the control experiments was monitored conventionally by determining the optical density at $\lambda = 540$ nm using a Spectronic 20+ spectrophotometer. Since this assay is not applicable for the NASA Bioreactor, bacterial cell growth had to be monitored via the dissolved oxygen using an Orion 850 DO meter or by monitoring the glucose consumption. The percentage saturation of oxygen in the medium was read with the probe that was calibrated with distilled water (101.7%). Dissolved oxygen in the bioreactor was measured by transferring the medium into a 100 mL beaker. A new fermentation had to be conducted after each measurement, as the medium was rendered non-sterile. The glucose content in the medium at different time points was determined by the use of a glucose oxidase assay (Trinder, Sigma Diagnostics). Samples of 1mL were collected from the shake flask and bioreactor using sterile disposable syringes. Cells were filtered off and the supernatant was used with the assay. All assays were performed in duplicate and results recorded using a Spectronic 20+ spectrophotometer. Dry cell mass (biomass) was determined gravimetrically. The cells were harvested by centrifugation (8000 rpm, 20 min), lyophilized and refluxed with methylene chloride or chloroform for 24 hours. After filtering the cells, the solvent was concentrated in a roto-evaporator and the polymer precipitated into cold methanol, dried and weighed. Structure determination was accomplished by ^{1}H NMR using a Bruker 400 MHz instrument. Chloroform-d_6 was used as solvent and solutions were adjusted to 0.5 % (w/v). Chemical shifts were recorded in parts per million (ppm) downfield from 0.00 ppm using tetramethylsilane as internal reference.

Results and Discussion

Poly(β-hydroxybutyrate), a biodegradable polyester has been synthesized for the first time under simulated microgravity conditions by fermentation of the bacterial strains *Azotobacter vinelandii UWD* and *Alcaligenes latus.*

Bacterial polymer production in the bioreactor differs from that in conventional shake flask experiments in the following aspects:

- Altered cell growth behavior
- No or negligible lag time
- Immediate polymer production
- Restricted polymer formation at extended growth times

The term bioreactor is used from here on to indicate experimental conditions that simulate microgravity, that is, constant state of free fall due to randomization of the gravity vector, allowing unrestricted, three-dimensional cell growth. Other aspects of true microgravity, as for instance exposure to elevated levels of radiation, were not realized. Bacterial growth in the bioreactor differs significantly from conventional shake flask experiments. Cells tended to grow in clusters, several small clusters became first visible after about 3 hours. Over a time period of 21 hours these clusters agglomerated into one cluster that was suspended in the center of the Bioreactor. By gradually adjusting the rotation speed of the bioreactor from initially 8 rpm to 31.5 rpm after 18 hours, the cell clusters were maintained in the center of the bioreactor, preventing them from colliding with the reactor wall. Figure 2 shows the progress of such a cluster formation.

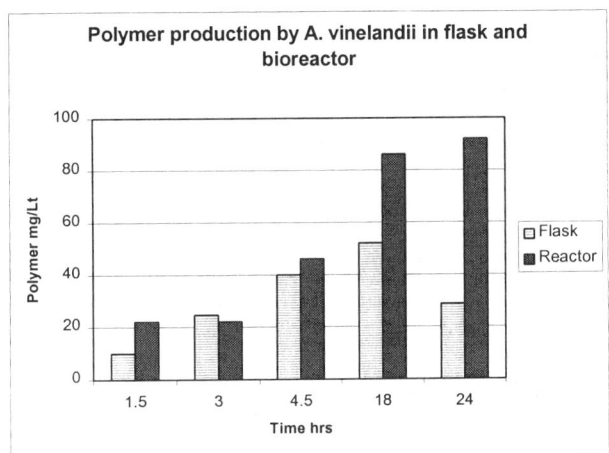

Figure 3: Comparison of polymer production by A. vinelandii UWD grown on minimum medium in shake flasks and in the NASA Bioreactor
(Reproduced with permission from *Polymer Preprints*, **2000**, *41(4)*, 1064 .)

All experiments in the bioreactor were compared to conventional shake flask experiments. Whereas the bacterial growth in the shake flask can be easily monitored by the optical density which is directly depended on the number of cells per volume unit, the monitoring of the cell density in the bioreactor was not possible, due to the cluster-like growth pattern, see Figure 2. Other analysis techniques had to be explored to reliably monitor the cell growth in the bioreactor. Two analysis techniques were determined to be suited for the evaluation of bacterial activity: the consumption of glucose and the oxygen profile.

The aeration in the bioreactor differs significantly from that in the shake flask. Aeration in shake flasks was accomplished by gas-exchange at the fermentation-broth-air surface area that was generated by the vigorous shaking motion. The bioreactor was aerated by a pump that delivered the moistened gas air bubble-free via a membrane to the reactor. All transport processes in the bioreactor were based on diffusion.

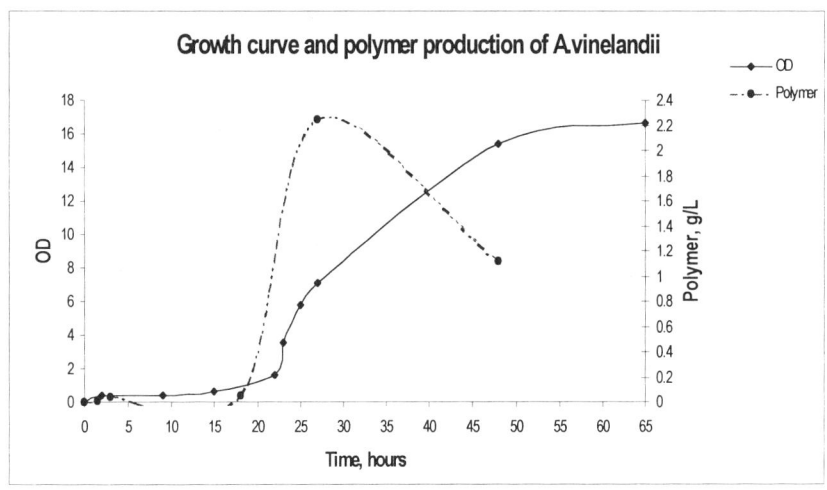

Figure 4: Growth of A. vinelandii grown on minimum medium and polymer yield

Bacterial fermentation in the Bioreactor led to a higher polymer production in the initial phase of the fermentation as compared to that achieved in shake-flasks, see Figure 3. After 24 hrs of fermentation, the polymer yield in the Bioreactor was about threefold of that obtained in shake flask experiments. At 24 hours 85 mg/L polymer were produced in the Bioreactor, whereas in the same time only 30 mg/L polymer were produced in the shake flasks. After 48 hours of fermentation however, the polymer production in the shake flask clearly superceded that in the bioreactor, 2.2 g/L in the shake flask as compared to 0.24 g/L in the bioreactor. The polymer production in the shake flask is so low in the first 24 hours since the cells are in their lag phase as it is clearly shown in the growth curve of *A. vinelandii*, see Figure 4. *A. vinelandii* exhibits a lag phase of about 20 hours with the polymer production starting shortly after the cells went into their exponential growth phase. Exhibiting a lag phase is a typical behavior for bacterial cells grown in conventional fermentation experiments.

In the initial stage of the fermentation, up to 24 hours, polymer production is more efficient in the bioreactor. Nutrients were transformed directly into the storage polymer, without extensive cell proliferation and the respective build-up of biomass, see Figure 5. As the fermentation proceeds, polymer production in the shake flask is eventually higher than that in the bioreactor. It is believed that the cluster-like growth pattern of the cells in simulated microgravity is the main reason for the eventual decrease in polymer production. This cluster-like growth pattern resulted in reduced mass transfer to and from the cell clusters. As the cells metabolized, they secreted waste products, mainly acetate, which accumulated in the center and immediate vicinity of the clusters. Since all mass transport processes were solely diffusion based, transport of the waste away from the cells and transport of nutrients to the cells was comparatively slow, probably going to the extent that the cells in the center of the clusters were eventually completely depleted of oxygen and nutrients. The build-up of metabolites in conjunction with reduced nutrient supply led to the observed lower polymer incorporation at a later stage of the fermentation in the Bioreactor. Most importantly, the lag phase described above for bacterial fermentation in conventional shake flask experiments was not observed when the cells were grown in simulated microgravity. As discussed above, the cells started immediately to produce polymer.

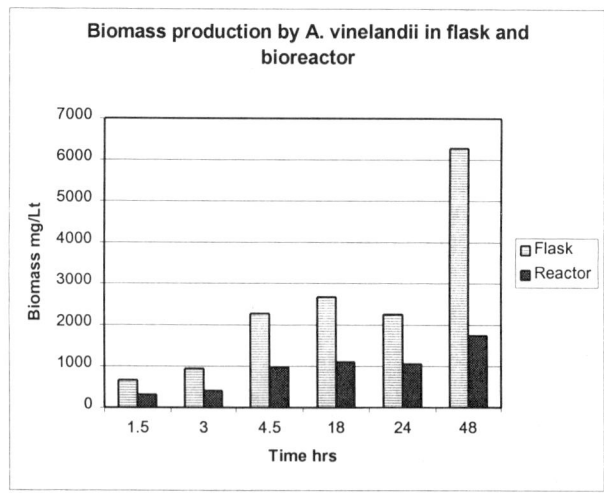

Figure 5: Comparison of biomass production by A. vinelandii UWD grown on minimum medium in shake flasks and in the NASA Bioreactor
(Reproduced with permission from *Polymer Preprints*, **2000**, *41(4)*, 1064 .)

In an effort to differentiate between the influence of simulated microgravity and differences in the oxygen supply to the cells, the oxygen profile in the bioreactor was adapted to match that of the shake flask as closely as possible. The established oxygen supply profile is given in Table 2. Since the initial phase of the bacterial growth is the one that exhibited significant differences compared to conventional shake flask experiments, the first 6 hours of the fermentation were monitored closely.

Table 2: Oxygen supply profile for the operation of the Bioreactor

Time Period [hr]	Oxygen supply [L/min]
0 – 1	1.36×10^{-2}
1 – 2	2.40×10^{-2}
2 – 6	$> 3.4 \times 10^{-2}$ [a]

[a] Oxygen was supplied in the excess of 3.4×10^{-2} L/min, a more accurate determination was not possible in the course of this experiment

Bacterial activity was monitored by recording the glucose consumption and oxygen profile, since the optical density could not be used to monitor the bacterial growth behavior due to the cluster like growth pattern. These experiments aimed at confirming or dismissing the hypothesis that bacterial growth is influenced by simulated microgravity. Hence, all fermentation parameters were kept identical to those in shake flask experiments. Figure 6 shows the consumption of glucose in the bioreactor and in the shake flask within the first four hours of the fermentation. Glucose was consumed in the bioreactor at a greater rate than in the shake flask experiments, indicating a higher bioactivity, that is, no lag phase was observed. The same tendency was observed by monitoring the dissolved oxygen for the first 6 hours of the fermentation, again oxygen was consumed at a much higher rate in the bioreactor than in the shake flasks, Figure 6. It was rather difficult to match the oxygen profile in the Bioreactor to that of the shake flask since no direct measurement of the dissolved oxygen was possible during a fermentation experiment. Supplying the Bioreactor with air resulted in a continuous decrease in the oxygen level (data not shown). Supplying the Bioreactor with the oxygen supply profile given in Table 2, led to the oxygen balance shown in Figure 6, matching the conditions in the shake flask up to 2 hours. Then, the oxygen level decreased due to the bioactivity of the cells. Independent of the amount of oxygen delivered to the fermentation broth in the Bioreactor, this distinct, steep decrease in oxygen levels was observed within one to three hours of inoculation and could not be counteracted by increasing the oxygen supply. This result indicated that biological activity started in the Bioreactor within one hour of inoculation, whereas in the shake flask experiments constant oxygen level between 85 and 90 % were maintained. Both, oxygen and glucose consumption indicated an elevated bioactivity of the bacterial cells in simulated microgravity. These results confirmed that the microorganisms did not undergo the conventionally observed, extended lag phase, when grown in microgravity. As shown above by the consumption of oxygen and glucose, bacterial cells grown in microgravity showed only a negligible lag phase, bacterial growth and polymer production started immediately after inoculation. In summary, the bacterial growth phenomena observed in the bioreactor can be attributed to the effect that microgravity has on the cells and not to an altered gas balance.

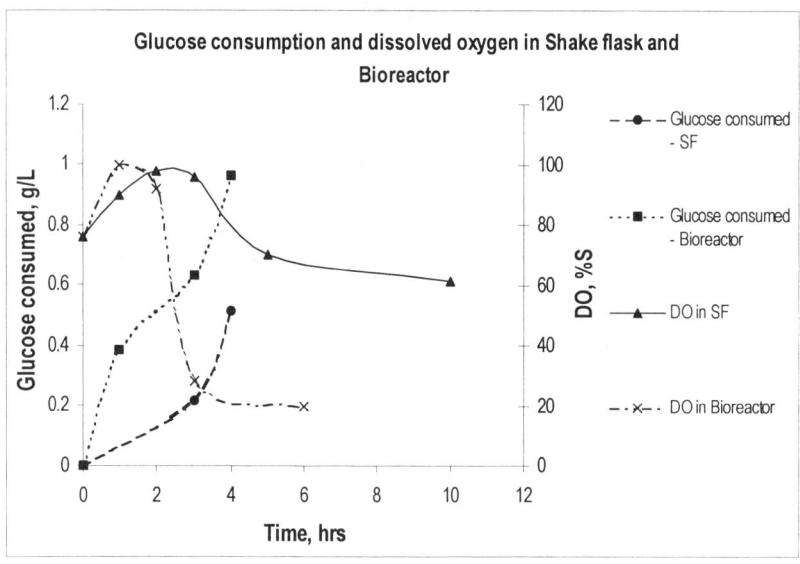

Figure 6: Glucose consumption and Dissolved Oxygen profile in the Bioreactor and in conventional shake flask experiments (SF)

In an effort to demonstrate that the effects of microgravity on bacterial polymer production are of general nature, initial experiments were conducted in the bioreactor using another bacterial strain, *Alcaligenes latus* (ATCC 29713). The growth and polymer production data confirmed the results obtained with *A. vinelandii*, that is, shortened lag time and immediate polymer production. The amount of polymer produced was higher than in the respective control shake flask experiments. Further investigations will aim at studying the growth behavior of this strain in simulated microgravity and under the condition of an adjusted oxygen profile.

Future work

Future research will concentrate on quantifying the microbial polymer production in simulated microgravity applying the established oxygen-profile. Initial results indicated that the polymer production in the bioreactor increased significantly when oxygen is supplied to the bioreactor. After three hours of fermentation, 110 mg/L biopolymer were obtained. This represented a threefold increase in polymer production in the bioreactor aerated with oxygen as compared to experiments when the bioreator was aerated with air only. Further detailing is necessary to reconfirm these preliminary data.

Acknowledgement

We thank Dr. Richard Ashby from USDA, Eastern Regional Devision for providing the *A. vinelandii UWD* bacterial strain. Financial support by NASA EPSCoR grant NCC5-391 and UAH Minigrant #3-87585 is greatly acknowledged.

References

1. Isoda, H.; Kitamoto, D. Shinmoto, H.; Matsumara, M.; Nakahara, T. *Biosci.Biotech.Biochem.* **1997**, 61(4), 609
2. Scholz, C. Mehta, S.; Nicolosi, R.; Bisht, K.; Guilmanov. V. Gross, R.A. *Polymer Preprints* **1998**, 39 (2), 168
3. Gutnick, D.L. Rosenberg, E. *United States Patent* 4,276,094
4. Gutnick, D.L. Nestaas, E.; Rosenberg, E.; Sar, N. *United States Patent* 4,883,757
5. Shine, J.; Dalgarno, L. Nature **1975**, 254, 34
6. Goeddel, D.V. in: Gene Expression Technology. Methods in Enzymology 185, New York Academic Press, **1990**
7. Klaus, D.; Simske, S.; Todd, P.; Stodieck, L. *Microbiology* **1997**, 143(2) 449
8. Mattoni, R.H.T. *BioScience* **1968**, 18, 602
9. Moatti, N.; Lapchine, L.; Gasset, G.; Richoilley, G.; Templier, J.; Tixador, R. *Naturwissenschaften* **1986**, 73, 413
10. Mennigmann, H.D.; Lange, M. *Naturwissenschaften* **1986**, 73, 415
11. Ciferri, O.; Tiboni, O.; DiPasquale, G.; Orlandoni, A.M.; Marchesi, M.L. .; *Naturwissenschaften* **1986**, 73, 418
12. Mattoni, R.H.; Ebersold, W.T.; Eiserling, F.A.; Keller, E.C.; Romig, W.R.; *The experiments of Biosatellite 2* (ed. J. Saunders) NASA SP **1971**, 104, 304
13. Planel, H. Tixador, R.; Nefedov, I.G.; Gretchko, G.; Richoilley, G. *Advances in Space Science* **1981**, 1, 95

14. Mennigmann, H.D.; Heise, M.; Fifth European Symposium on Life Science and Space Research, Arcachon, France, *Conference Proceedings* **1983**, 83
15. Mennigmann, H.D. *BioEngineering* **1988**, 4, 20
16. Mennigmann, H.D.; Lange, M. in: BIORACK on Spacelab D1 ESA SP-1091 (ed. N. Longdon, V. David) *ESA Publ. ESTEC*, Noordwijk, **1987**, 374
17. Gmünder, F.K.; Cogoli, A. *Appl. Microgravity Techn.* **1988**, 1, 115
18. Doi, Y. *Microbial Polyesters* VCH, New York **1990**
19. Steinbüchel, A. in: Biomaterials: Novel Materials from Biological Sources (ed. D. Byrom) Macmillan Publishers Ltd. London **1991**, 123
20. Huisman, G.W.; Wonink, E.; Meima, R.; Terpstra, P.; Witholt, B. *J.Biol.Chem.* **1991**, 266(4), 2191
21. 21 Marois, Z.; Zhang, Z.; Vert, M.;Deng, X.; Lenz, R.W.; Guidon, R. *J.Biomater. Sci. Polym. Ed.* **1999**, 10(4), 483
22. Scholz, C. in: *Polymers from Renewable Resources Biopolyesters and Biocatalysis* (ed. C.Scholz and R.A. Gross) ACS series 764, **2000**, 328

Chapter 15

Development of Impedance Spectroscopy to Study Polymerization Processes in Microgravity

A. P. Kennedy and J. McLendon

Department of Chemistry, Morgan State University, Baltimore, MD 21251

Because of the high cost of microgravity research, it is important to develop techniques that can continuously monitor the polymerization process in space. Impedance spectroscopy is one technique, which is being developed to complement other techniques used to study polymerization reactions. Impedance spectroscopy has been used successfully to study the photopolymerization of diacetylene both in solution and at the surface. There are distinct differences between the measurements of the solution and surface polymerization processes. In addition, the effect of convection on the polymerization process was observed using this technique. Impedance spectroscopy can potentially become a useful tool for monitoring polymerization processes in microgravity.

Introduction

In-situ processing of polymer structures in space will be necessary to build and repair the space station, as well as to develop radiation shielding on the moon and Mars (1). Only a few investigations have been performed on polymer reactions in a microgravity environment. The majority of microgravity investigations have involved biological and inorganic materials and processes.

Polymeric systems have not been investigated as extensively as other materials because polymers were considered too viscous for any microgravity effects such as convection or sedimentation to be important.(*2*) However, Briskman has recently reported that gravity can affect polymerization through several mechanisms.(*3*) He has shown that four mechanisms; sedimentation of polymeric microglobules, thermal and concentration convection, instability of polymerization front, and the effect of mass forces on a deformable polymer can influence polymerization processes and polymer structures. Determining the importance of these mechanisms during polymerization will help advance our understanding of polymerization both on earth and in space.

In order to understand the effects of these gravitational mechanisms in microgravity, the polymerization process has to be monitored throughout the duration of the experiment. At the present time the majority of polymerization experiments conducted in microgravity are not monitored during flight. In-situ monitoring techniques can make these experiments more cost effective and provide meaningful real time data. Development of in-situ monitoring techniques will also be important to insure the integrity of the structures fabricated in space. Many different techniques have been used to investigate polymerization processes on earth, however impedance or dielectric spectroscopy provides a simple yet powerful technique for following the complete polymerization process from monomer through fully cured polymer. Impedance measurements are performed usually by holding either the temperature or the frequency constant. Under these conditions changes in impedance are associated with molecular motions. The high frequency responses are due to alpha and/or glass transitions while other molecular motions such as side group rotations and ionic diffusion occur at lower frequencies. The resulting response has been correlated with the well-established Debye theory. These molecular motions are related to changes that occur for a fully polymerized system. Dielectric spectroscopy has been used to study the polymerization of a variety of thermoset resins, however most research has been conducted on epoxy thermosets. The dielectric behavior of epoxy resins can be adequately described by the Debye model, in which the dielectric constant, ε', is given by:

$$\varepsilon' = \varepsilon_u + \frac{(\varepsilon_r - \varepsilon_u)}{1 + (\omega\tau)^2} \qquad (1)$$

where ε_u and ε_r are the unrelaxed and the relaxed dielectric constants, respectively, ω is the angular frequency given by $\omega = 2\pi f$, and τ is the characteristic relaxation time. The dielectric loss ε'' is given by:

$$\varepsilon'' = \frac{\sigma}{\omega \varepsilon_o} + \frac{(\varepsilon_r - \varepsilon_u)\omega\tau}{1+(\omega\tau)^2} \qquad (2)$$

where σ is the ionic conductivity, and ε_0 is the permittivity of free space. The first term represents the ionic conductivity of the material and becomes dominant at low frequencies. This term will vary with the viscosity of the material and the ionic concentration, which can also change during cure. The second term is the dipolar component of the dielectric loss. This term changes as the network forms and as the viscosity changes. A maximum is observed in the dielectric loss spectra when $\omega\tau=1$, and experimentally the relaxation time can be determined from the loss peak. In the case of a curing resin, ε_r and ε_u would correspond to the neat resin and cured polymer, respectively. This model is based on a static number of dipoles that respond to variations in the applied frequency. Although of the dipoles in the monomer and curing agent are consumed during the polymerization process, the model has worked well for epoxy resins.

Impedance spectroscopy has been used to study the polymerization of variety of thermoset resins, but only a few investigations have been done on solution polymerization reactions.(4,5) We will present one of the first impedance measurements that monitors both the surface and bulk photopolymerization of 6-(2-methyl-4-nitroanilino)-2,4-hexadiyn-1-ol (DAMNA) (Figure 1). Frazier et al (6,7) reported the surface polymerization of polydiacetylene films from the photopolymerization of DAMNA monomer solutions. This unique photopolymerization process produces amorphous thin films of polydiacetylene, which can be used in the fabrication of waveguides and photonic devices. However films formed on earth are low quality and have defects that appear to be the result of buoyancy-driven convection. Microgravity experiments on the same system produced very high quality optical films. The lack of in-situ measurements limited the analysis of the effect of gravitational forces on the development of defects in the poyldiacetylene films.

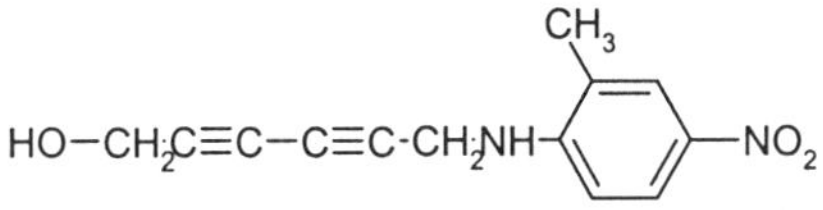

HO−CH$_2$C≡C−C≡C-CH$_2$NH−⟨C$_6$H$_3$(CH$_3$)⟩−NO$_2$

Figure 1. DAMNA monomer

Experimental

The impedance measurements were made using a HP4274A LCR meter. Measurements were taken at 400 Hz,1 , 2, 4, 10, 20, 40, and 100 kHz. The LCR meter was interfaced to the PC using Labview software. The program allowed measurements to be taken at 30 sec intervals or longer for the duration of the run. Gold plated quartz interdigitated electrodes were purchased from DekDyne and used to measure the impedance of the solutions. A 100W UV lamp was used at a distance of 10 cm from the cell. Data was taken in 5 minute intervals for at least 24 hours. Solutions containing 2.5 mg/mL of DAMNA in 1,2 dichloroethane were used.

A temperature controlled cell was designed to permit irradiation from both sides. A 2.54 cm quartz window was placed on one side and the interdigitated electrode on the other. Solution polymerization was monitored by irradiating through the quartz window while the electrode was in contact with the solution on the opposite side (Figures 2a and 3a). It should be noted that when monitoring solution polymerization only processes occurring near the surface of the electrode could be monitored. Polymerization at the surface was monitored by measuring the impedance while irradiating the solution through the back of the electrode (Figures 2b and 3b). Because of the differences in transmission characteristics of the quartz window and the electrode, quantitative comparisons could not be made. The dielectric properties could not be determined because of the irregular surface area that was irradiated.

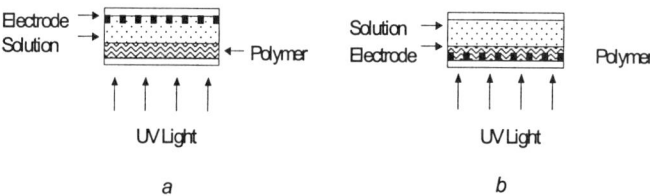

Figure 2 a & b. Bulk and Surface polymerization with convection

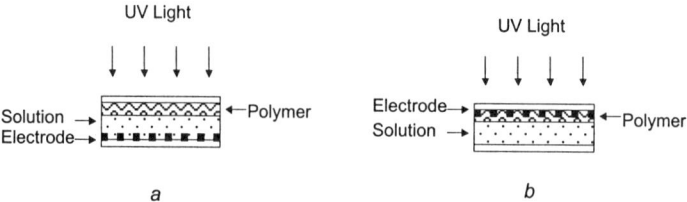

Figure 3 a & b. Bulk and Surface polymerization without convection

Results and Discussion

A comparison of the surface and bulk polymerizations measured at 1 kHz are shown in Figures 4 and 5. There is a distinct difference in both the capacitance and the dissipation factor between the surface and bulk polymerizations at the lower frequencies. When the lamp was turned on there was an initial decrease in the capacitance (Figure 4) and an increase in the dissipation factor for the surface polymerization (Figure 5). This was due to the heating of the electrode by the UV lamp. A similar decrease was observed when an empty cell was illuminated. There is a rapid increase in the capacitance for the surface polymerization that indicates the responses are due to the deposition of polydiacetylene on the surface by the photopolymerization reaction. There is a peak in the dissipation factor at 120 minutes during the surface polymerization. A maximum is observed in the case of the bulk polymerization but it decreases slowly with time. The capacitance of the surface polymerization is a factor of four larger than the bulk. This clearly indicates that polymer film is growing on the surface of the electrode. The capacitance increases rapidly during the first 500 minutes and then begins to plateau at longer times. Paley et.al. (8) studied the kinetics of the thin film growth for polydiacetylene and reported the rate to be first order in intensity and half-order in monomer concentration. The estimated thickness of the film after 500 minutes was 1 micron. It is reasonable to assume that after 500 minutes the film is opaque to the UV radiation and film growth decreased with a resulting reduction in capacitance. A similar trend was observed for the dissipation factor.

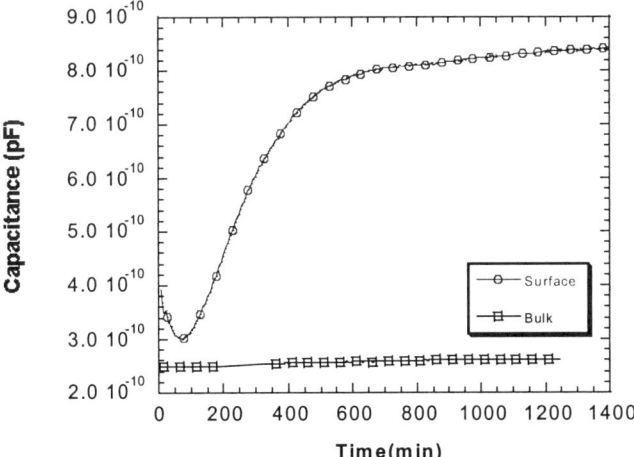

Figure 4. Capacitance for surface and bulk solution at 1 kHz

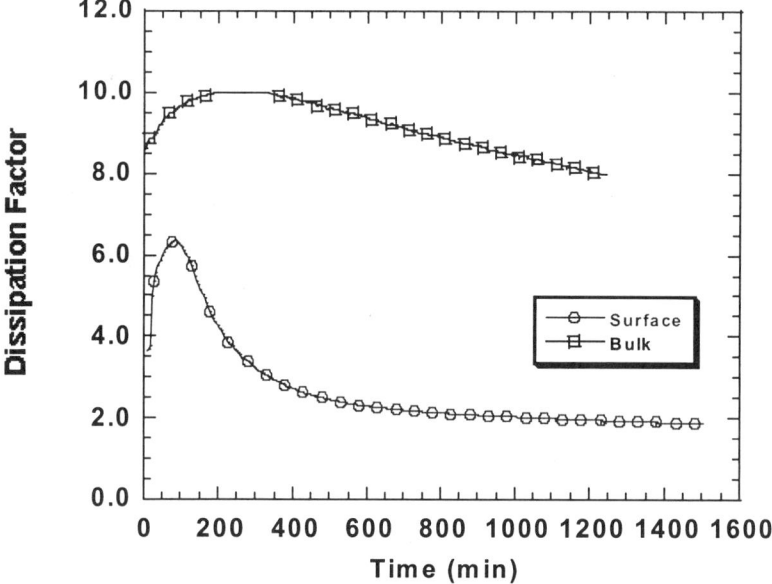

Figure 5. Dissipation factor surface and bulk solution polymerization at 1kHz

To determine if convective effects can be observed using this technique the sample cell was placed in various orientations. Figures 6 and 7 show the comparison in surface polymerization irradiated from above and below when the sample is in the horizontal orientation. (See Figures 2b and 3b.) The capacitances for both measurements are identical except for differences at longer times. This suggests that gravitational effects do not influence surface polymerization at early times. However when convection is dominant (irradiated from bottom) the capacitance does decrease at longer times. Since the reaction has slowed at this point, the decrease in capacitance is probably due to the removal of polymer from the surface by convection.

Figure 8 and 9 show the results for the solution polymerization at various orientations. There is a distinct difference between irradiating from above and below in this case. (See Figures 2a and 3a.) Gravitational effects have a more pronounced influence on polymerization in solution than at the surface. When irradiated from below convection is dominant and capacitance increased. The increase in capacitance is more gradual and changes less than the surface polymerization. When irradiated from above convection is minimized and there is no change in the capacitance. This suggests that the polymer is formed at or near the surface, and no significant amount of polymer is transported to the bottom of the solution when convection is suppressed. This also suggests that sedimentation of insoluble crystallites is not significant in this orientation. However, when convection is dominant it is possible to form a significant concentration of oligomers in solution, which could form defects in the thin film surface.

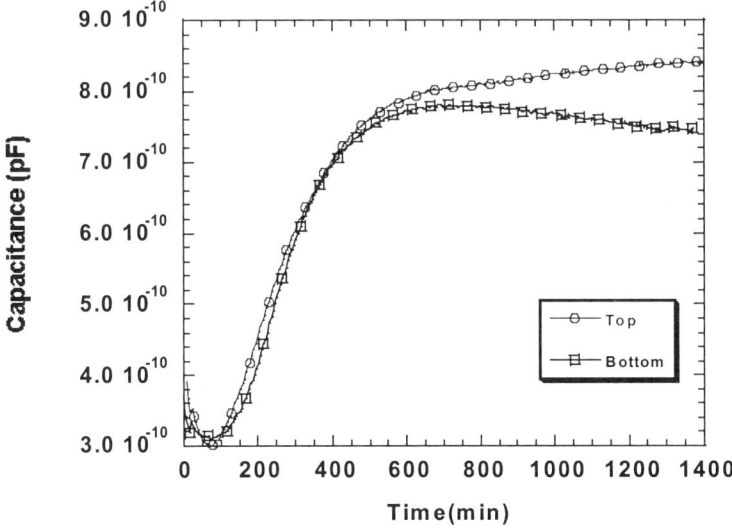

Figure 6. Capacitance for the surface polymerization of 2.5 mg/mL of polydiacetylene in horizontal orientation.

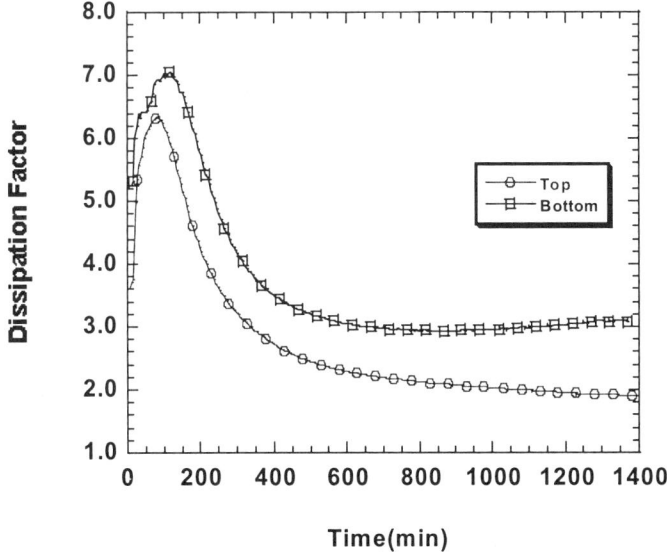

Figure 7. Dissipation factor for the surface polymerization of 2.5 mg/mL of polydiacetylene in horizontal orientation.

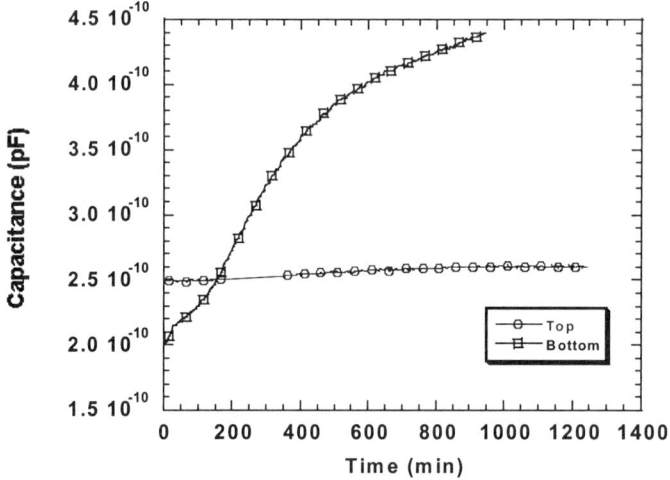

Figure 8. Capacitance for the bulk solution polymerization of 2.5 mg/mL of polydiacetylene in various orientations.

Figure 10 shows the change in capacitance vs. frequency. Both the capacitance and the dissipation factor decrease monotonically with increasing frequency. This frequency dependence is indicative of ionic conductivity. The increase in capacitance as a function of time at the lower frequencies appears to be due to the formation of a thin film on the surface of the electrodes.

Conclusions

The in-situ impedance spectra for the surface and bulk solution polymerizations of polydiacetylene have been successfully measured. The results show a distinct difference between the response of the surface and bulk polymerization. Surface polymerization is not affected by convection except at longer times when the polymer film appears to be decreasing in thickness. This suggests that photopolymerization is occurring primarily at the surface and polymerization in solution does not contribute significantly to film growth. Convective effects have a strong influence on bulk polymerization. When convection is minimized the reaction products appear to be localized near the surface. There is no evidence that sedimentation of insoluble polymers occurs. Under the present experimental conditions, this technique can only monitor film growth. Future studies will determine whether the kinetics of the film growth can be obtained from this technique. In addition, this technique will be used to monitor a variety of other important polymerization processes in an effort to further access the utility of this technique in monitoring microgravity polymerization experiments.

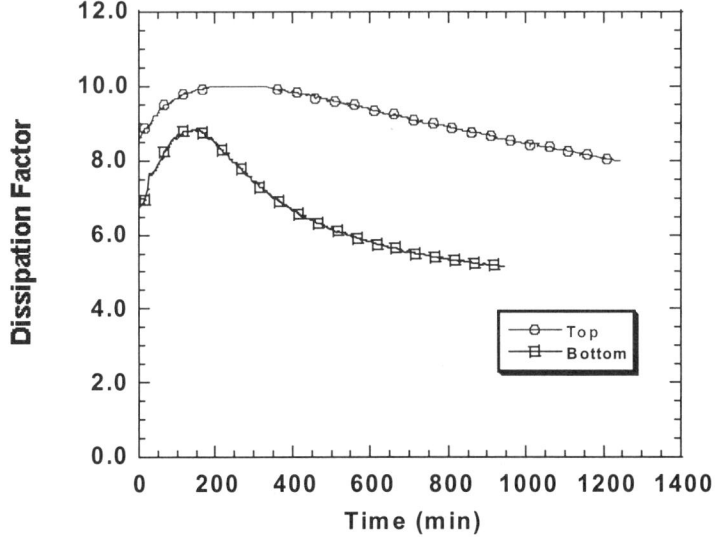

Figure 9. Dissipation factor for the bulk solution polymerization of 2.5 mg/mL of polydiacetylene in various orientations

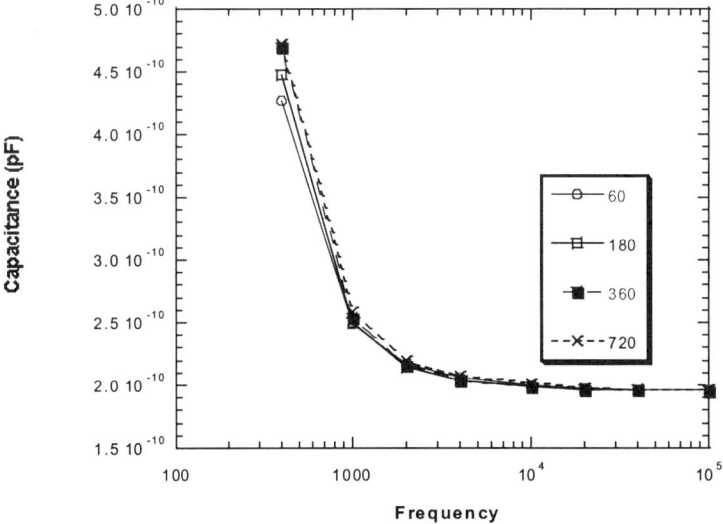

Figure 10. Capacitance vs. Frequency at various times (minutes).

Acknowledgment.

A special thanks to Drs. Don Frazier and Steve Paley of NASA Marshall Space Flight Center.

References

1. Kim MHY, Thibeault SA, Wilson JW, Simonsen LC, Heilbronn L, Chang K,Kiefer RL, Weakley JA, Maahs HG, *High Performance Polymers*, **2000**, 12: (1), 13-26
2. *NASA Conference Publication;* D. O. Frazier, C. E. Moore (1993); 3250.
3. Briskman V.A., *Gravitational Effects in Materials and Fluid Sciences*,**1999**; 24: (10), 1199-1210.
4. L. Matejka, K. Dusek, *J. Poly. Sci.:Part A*, **1995**, 33, 461
5. L. Matejka, S. Podzimek, K. Dusek, *J. Poly. Sci.:Part A*, **1995**, 33, 473
6. H. Abdeldayem, D. Frazier, M. Paley, B. Penn W. Witherow, C. Bank, A. Shields, *SPIE*, **1996**, 2809, 126.
7. D. O. Frazier; R.J. Hung; M.S. Paley; Y.T. Long; *J. Crystal Growth,* , **1997**, 175, 172.
8. M.S. Paley, D. O. Frazier,S.P. McManus, S.E. Zutaut, M. Sanghadasa *Chem Mater*, **1993**, 5(11), 1641-1644

Chapter 16

Gravity Effects during Mold Filling: Fluid Front Dynamics

M. C. Altan and K. A. Olivero

School of Aerospace and Mechanical Engineering, University of Oklahoma, Norman, OK 73019

Gravity effects on free surface flows during mold filling are experimentally studied and found to be significant in some flow regimes. An experimental disk-shaped mold cavity is constructed which allows for flow observation and measurement of spreading, the tendency for the bottom of the flow front to advance ahead of the top due to gravity. A method is developed in order to perform experiments to investigate the effect of pertinent non-dimensional parameters. Results are presented at three radial cavity locations isolating each of these non-dimensional parameters. Significant dependence of spreading on Bond number is observed.

Molding operations involving low viscosity polymers such as reaction injection molding (RIM) are commonly used to manufacture net-shape components quickly and inexpensively. These molding operations typically involve the displacement of air in a cavity by a polymeric resin, which is either cooled or cured to form the final solid part. The interface between the polymer and the air, referred to as the free surface, progresses through the cavity during filling and may be influenced by a variety of factors including viscous forces, surface forces, and gravity. In most reaction injection molding, curing starts during the filling and is usually completed within a minute. Therefore, the dynamics of the free surface could affect heat transfer as well as curing rate in the mold, and thus ultimately the mechanical properties of the molded part. Non-uniform curing can result in a variety of defects, including non-homogenous mechanical properties, warping and void formation.

A number of experimental and theoretical studies dealing with free surfaces are published in literature. The most common approach in determining free surface shapes with respect to molding processes is to assume the Capillary number is high, and thus viscous forces dominate surface forces. For example, in Behrens et al. (1), the fluid front shape is completely dictated by the flow kinematics and gravity effects are not considered. Other studies address the issue of the viscous stress singularity and so-called slip length used in analysis of moving contact lines (2,3). However, these do not address the free surface shape over the entirety of the free surface. Closely related are investigations to establish dynamic contact angles (4-6). In these studies, the focus is also on the flow dynamics near the contact point. Blake investigated gravity effects on free surface shape in mold filling (7,8); however this study involves a vertically aligned cavity in which gravity does not cause an asymmetric fluid front. Several studies have been performed which include gravity effects on the spreading of a liquid drop on a solid surface including gravity effects on the shape and motion of the free surface (e.g., Chen, et al. (9) and Hocking (10)). These studies tend to involve slow-moving, quasi-static free surfaces, and may not be directly applicable to forced flow in a mold cavity.

A trend in molding processes is towards the use of lower viscosity resins. These resins allow for higher fill rates and lower injection pressures, thus leading to higher production rates and lower equipment cost. However, the use of low-viscosity polymers increases the complexity of flow dynamics. For example, Reynolds numbers governing the flow may become too large for Stokes flow assumptions to be valid. In addition, gravity effects on the free surface, which are usually neglected, can have pronounced effects on these flows, particularly on filling patterns and residence times. For planar mold cavities the flow is typically assumed to be symmetric about the midplane. However, if gravity effects become important, the bottom of the flow front can advance well ahead of the top. In such cases, if the gravity effects are neglected in numerical mold filling simulations, the shape and location of the weld lines (i.e., the lines at which the fluid front meet during filling) may be inaccurately predicted. Example fluid front shapes with and without sagging due to gravity are depicted in figure 1.

The aim of the current study is to experimentally quantify the effect of gravity on the free surface shape during the filling of a mold cavity. It is of particular interest to isolate the important non-dimensional parameters governing the fluid front dynamics. Towards this end, mold filling experiments are performed to characterize the spreading of the fluid front during filling of a disk-shaped cavity. The independent effects of each of the Reynolds, Bond, and Capillary numbers on spreading are observed.

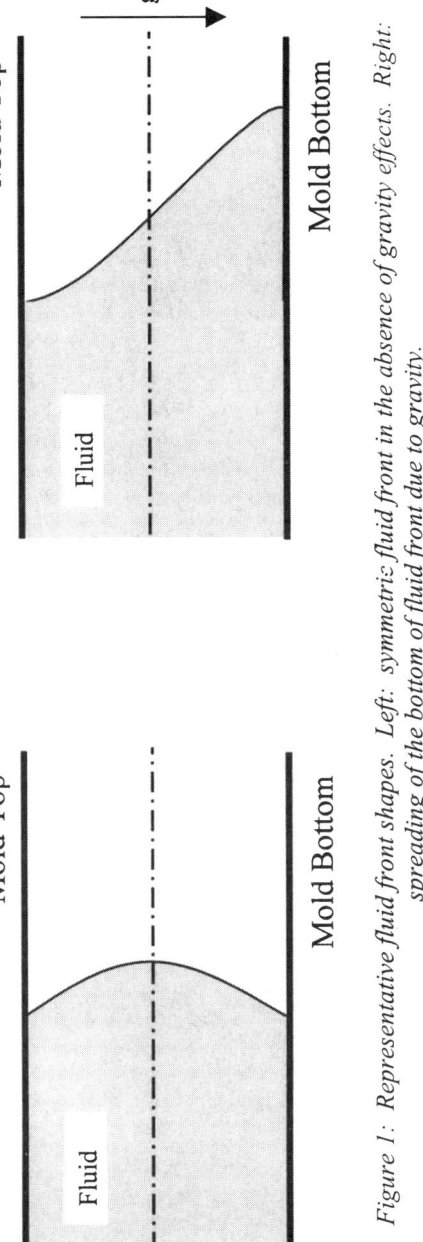

Figure 1: Representative fluid front shapes. Left: symmetric fluid front in the absence of gravity effects. Right: spreading of the bottom of fluid front due to gravity.

Experimental Setup

An experimental molding setup is constructed to observe and measure the features induced by gravity during the filling of a mold cavity. The setup consists of a disk-shaped mold cavity containing various embedded sensors, a peristaltic pump to inject the fluid, and a PC based data acquisition system for monitoring the flow. The mold cavity is formed by placing aluminum spacer plates between one-inch-thick Plexiglas sheets. A circular section with a 9-inch radius is cut out from the center of each spacer plate. Various gap-widths ranging from 0.0625 to 1.0 inch are attainable by using different combinations of these plates. Shim steel spacers having thicknesses 0.004, 0.008, and 0.012 inch are inserted between the spacer plates and Plexiglas to achieve more precise gap-width control. Inlet gate diameters of 0.125, 0.250, and 0.375 in. can be selected by inserting one of three available inlet gates fabricated from Plexiglas. In addition, the mold can be turned over to allow the inlet to be placed on either the top or bottom mold wall. Pressure transducers and fluid front sensors for flow measurements are mounted in both the top and bottom Plexiglas sheets along radial lines as illustrated in figure 2. The advancing fluid front is sensitive to the surface quality of the mold walls, thus, the Plexiglas mold walls are sanded at the sensor locations by a series of 400, 600, 1000, and 4000 grit sandpaper, followed by polishing with a buffing wheel and polishing compound. This greatly reduces perturbations to the fluid front shape as the front passes the sensor locations.

Glycerol diluted with water is selected as the filling fluid. Fluid properties can be varied easily by changing the volume fraction of water in the mixture. The mixture behaves as a Newtonian fluid and viscosities ranging from 1 to 1200 cP can be obtained. Most experiments in this study utilize viscosities between 50 and 400 cP, which are similar to the viscosities observed during reaction injection molding process (*11*). It should be noted that non-Newtonian effects are shown to be small for some polymers used in reaction injection molding (*11*). The fluid is injected at constant flow rate by a Masterflex IP73 microprocessor controlled peristaltic pump. Two pump heads can be configured in a parallel arrangement to yield flow rates for water ranging from 4×10^{-6} m^3/s to 2.7×10^{-4} m^3/s at a maximum pressure of 40 psi. For higher viscosity fluids, the flow rates are somewhat reduced. The pump and controller are labeled on the photograph of the experimental setup shown in figure 3. An inline pulse dampener, not visible in figure 3, minimizes flow fluctuations which are common to peristaltic pumps.

Figure 4 is a close up diagram of one of the three fluid front sensors mounted in the mold walls at $R=2$, 4, and 6 in. Each sensor consists of one power terminal and three sensing terminals as shown in figure 4. A sensing

231

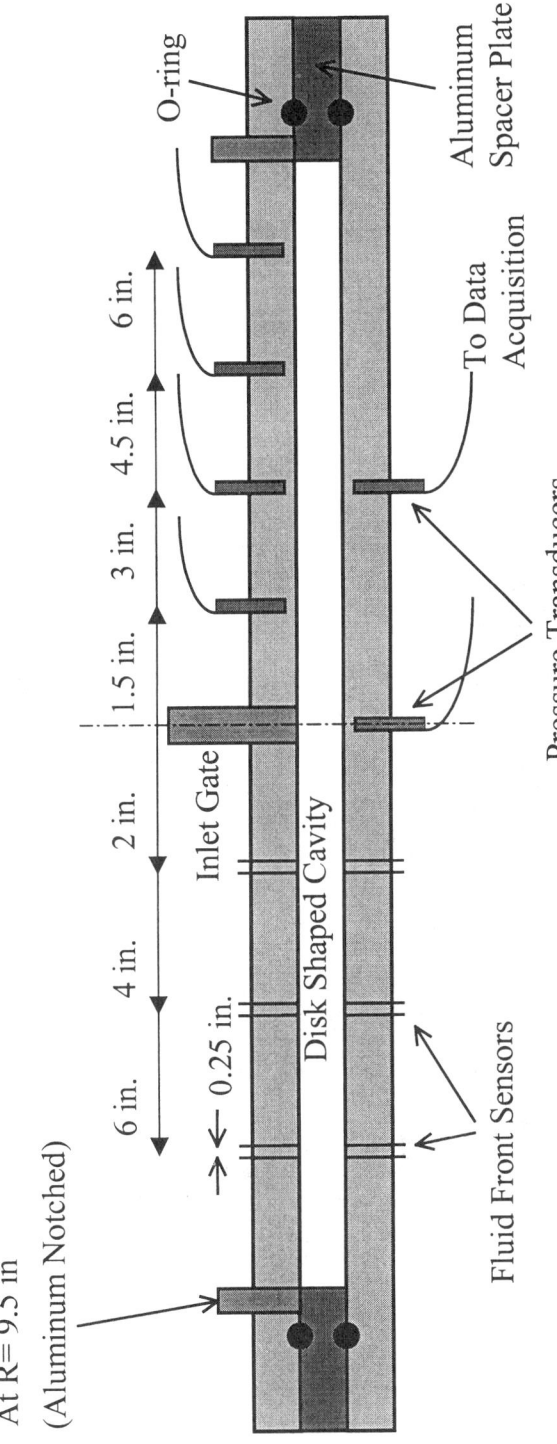

Figure 2: Experimental mold cavity diagram: cross section and dimensions.

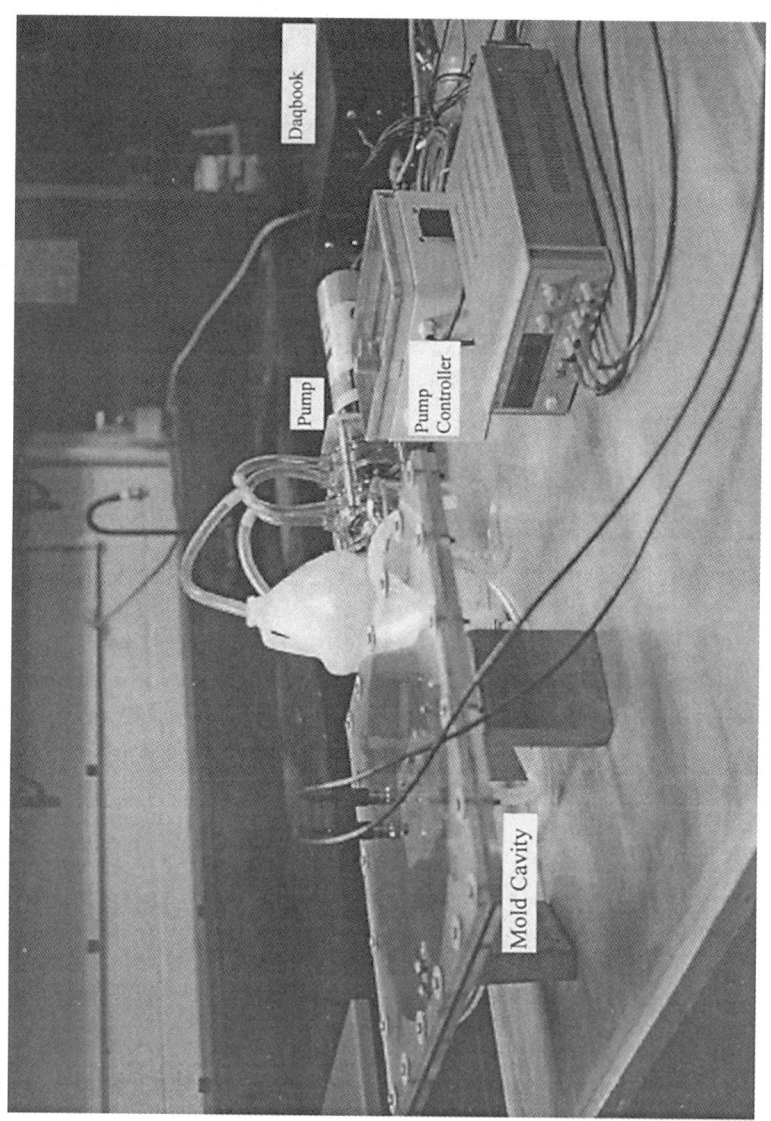

Figure 3: Setup for mold filling experiments. The mold is shown with three pairs of fluid front sensors and two pressure transducers placed on the top wall.

233

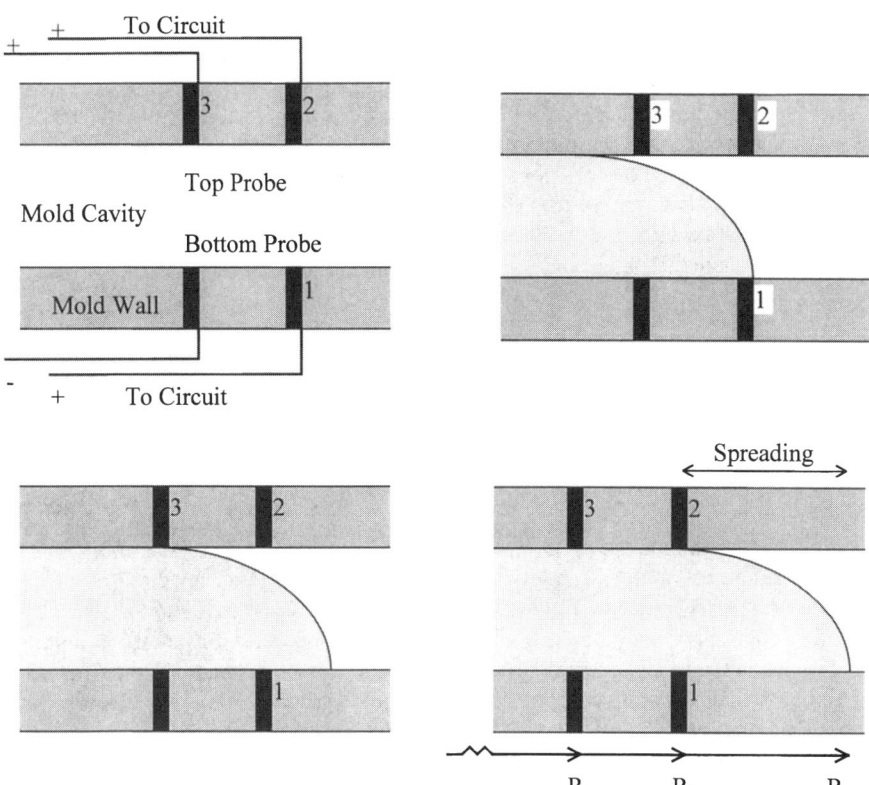

Figure 4: Schematic cross section of probe wiring depicting the advancement of fluid front through a probe location.

circuit connected to a data acquisition system measures contact times based on the corresponding resistance drop when the fluid contacts each of the three sensing terminals. The circuit operates by monitoring the voltage across a reference resistor in series with the sensing terminal. A change in voltage indicates a resistance change across one of the sensing terminals due to fluid contact. An op amp in the circuit amplifies the voltage reading and conditions the signal, significantly lowering its output impedance (the data acquisition system has a maximum input impedance of 1000 ohms.) A total of nine such circuits are attached to each of the sensing terminals. The three contact times are clearly identifiable in the raw voltage data obtained from each of the three fluid front probes.

The spreading at each radial location is calculated from the times at which the fluid front contacts terminals 1, 2 and 3 of a given fluid front probe (i.e., t_1, t_2, and t_3), and the two radial locations of these terminals, R_a and R_b (all labeled in figure 4) It is assumed that each point on the fluid front travels at uniform velocity at a given radius. The equation for spreading is derived in ref. (*12*) based on the contact times and terminal radii as,

$$S = \sqrt{R_b^2 + \frac{(R_b^2 - R_a^2)(t_2 - t_1)}{(t_2 - t_3)}} - R_b. \tag{1}$$

Isolating Non-Dimensional Parameters

The experimental setup is designed to simulate mold filling operations over a wide range of flow parameters. Dependence of spreading on these physical filling parameters, such as flow rate and gap-width, can be easily ascertained by varying the individual parameters independently. However, it is useful to obtain the dependence of flow dynamics on the pertinent non-dimensional parameters. This is difficult to achieve by arbitrarily selected fill conditions, as changing a single parameter, such as the volume flow rate, can result in different values for all the non-dimensional parameters.

There are three non-dimensional parameters expected to affect spreading and flow characteristics during an isothermal, Newtonian filling: the Reynolds, Bond, and Capillary numbers. A fourth non-dimensional value, the contact angle, is a characteristic of the experimental apparatus (i.e., the contact angle depends on the fluid injected and the mold surface.) Thus, a method needs to be developed to determine the experimental parameters; i.e., the gap-width, L, the volume flow rate, Q, and the properties of the filling fluid, to achieve a prescribed set of non-dimensional parameters. To obtain this solution, the Reynolds, Bond, and Capillary numbers are defined as

$$\text{Re} = \frac{\rho u L}{\mu}; \quad Bo = \frac{\rho g L^2}{\sigma}; \quad Ca = \frac{\mu u}{\sigma}, \qquad (2)$$

where g is the acceleration due to gravity, and ρ, σ, and μ are the fluid's density, surface tension, and viscosity, respectively. It is clear from eqs 2 that as gravity goes to zero, Bond number will be zero and the other non-dimensional parameters, Re and Ca, remain unaffected. Thus, experiments in microgravity could be utilized to further investigate the effects of Reynolds and Capillary numbers in the absence of gravity (i.e., $Bo=0$).

Non-dimensional parameters are defined at the location of the first fluid front probe ($R=2.0$ in.), and thus the velocity, u, represents the average velocity at $R=2.0$ in. The velocity, u, scales linearly with the volume flow rate, Q, and represents a characteristic velocity scale. These expressions are rearranged and combined to eliminate two of the physical experimental parameters, u and L as,

$$0 = \left(\frac{\rho Bo}{g}\right)^{1/4} \left(\frac{Ca}{\text{Re}}\right)^{1/2} (\sigma)^{3/4} - \mu. \qquad (3)$$

In eq 3, gravity is a known constant and Re, Ca, and Bo are constant parameters, leaving three unknowns. It is observed that all unknown parameters in eq 3 (i.e., μ, ρ, and σ) are properties of the filling fluid. Glycerol diluted with distilled water is selected as the filling fluid, and the fluid properties are obtained experimentally based on the volume fraction of glycerol in the mixture, f, as,

$$\rho = \rho_w(1-f) + \rho_g f, \qquad (4)$$
$$\sigma = 72.3 - 8.3f,$$
$$\mu = 134.67f^3 - 60.238f^2 + 8.4509f + 1$$
$$0 < f < 0.6875$$
$$\mu = -53268f^5 + 411091f^4 - 934784f^3 + 940459f^2 - 442393f + 79755$$
$$0.6875 < f < 1.0,$$

where the subscripts w and g refer to water and glycerol respectively. These fluid properties are also dependent on the temperature of the mixture, thus to avoid temperature effects all measurements and experiments are performed at 25°C. It is determined that all three fluid properties can be expressed as a function of the volume fraction glycerol in the mixture, f, and eq 3 is expressed with a single unknown as,

$$0 = \left(\frac{\rho(f)Bo}{g}\right)^{1/4} \left(\frac{Ca}{Re}\right)^{1/2} (\sigma(f))^{3/4} - \mu(f). \tag{5}$$

To perform an experiment, the desired set of non-dimensional parameters, Re, Ca, and Bo are initially selected. These parameters are used in eq 5 to determine the required f numerically. After obtaining the necessary f value, the remaining independent physical parameters, L and u, are determined using eqs 2. Thus all experimental parameters are uniquely determined for the selected set of non-dimensional parameters, Re, Bo, and Ca.

Results

Eleven sets of mold filling experiments are performed at various Bond and Reynolds numbers. In each set, spreading is measured at 2, 4, and 6 inches away from the inlet gate as described earlier. In all these eleven sets, the Capillary number is kept constant at 0.02. Additional experiments can be performed to determine the effect of Capillary number on spreading as Re and Bo are kept constant. Each filling experiment is repeated between 4 to 8 times, and the measured spreading values are given with a 95% confidence interval. Table I shows the selected Bond and Reynolds numbers of the eleven sets of mold filling experiments. Experiments are also performed at *Bo*=12.75; however, at this Bond number, the bottom of the fluid front exits the mold cavity before any of the top sensors are contacted, thus resulting in "infinite" spreading. The experimental setup is capable of achieving higher Reynolds numbers at higher Bond number, but Re=4.0 is the maximum at which reliable data can be collected. The lower two Bond numbers have an upper Reynolds number limit of 2.0 due to flow rate limitations of the pump for the higher viscosities.

Table I. Bond and Reynolds numbers used for eleven sets of mold filling experiments.

Bo=4.35	*Bo*=5.75	*Bo*=8.20
Re=0.04	0.04	0.04
1.0	1.0	1.0
2.0	2.0	2.0
		3.0
		4.0

Spreading results from the eleven sets of experiments are depicted in figures 5-7. Each figure represents spreading as a function of Reynolds number at one of the three radii, $R=2$, 4, and 6 in. respectively. Three spreading curves are contained on each figure for the three Bond numbers in Table I. In all curves, spreading is found to vary linearly with Reynolds number, with very little change in spreading over two orders of magnitude increase in Reynolds number. Bond number, however, is found to influence spreading considerably. At all radii, up to one order of magnitude increase in spreading is shown with increasing Bond number. Experimental error is also observed to increase at higher Bond numbers, due to larger spreading values. Increased flow rate, and thus higher fluid front velocities, contribute to higher errors at increased Reynolds numbers as well.

In figure 5 ($R=2$ in.), a maximum spreading of about 0.5 in. is observed at $Bo=8.20$. In figures 6 and 7, the maximum spreading at $Bo=8.20$ is observed to increase significantly with increasing radius to about 1.0 in. at $R=6$ in. However, at the lower Bond numbers $Bo=4.35$ and 5.75, spreading is found to vary only slightly with increasing radius, with the highest spreading occurring at $R=4$ in. at all Reynolds numbers.

Figures 8 and 9 depict an alternate method of observing the significance of spreading for different fill parameters. These figures represent a cross section of the mold cavity, with the inlet indicated on the x-axis by $R=0$ and the flow in the R direction. The z-axis represents the direction perpendicular to the flow plane. For different experiments the specified gap-width varies by as much as 10%, however all profiles are drawn using the average of 0.25 in. for comparison purposes. Linear flow profiles are drawn using the spreading data at each of the three fluid front probe locations (i.e., $R=2$, 4, and 6 in.) In figure 8, profiles are drawn at three Reynolds numbers, with the Bond and Capillary numbers held at 5.75 and 0.02, respectively. It is observed that spreading is on the order of the gap-width, and unaffected by varying Re. Figure 9 depicts spreading at three Bond numbers, with the Reynolds and Capillary numbers held constant at 0.04 and 0.02, respectively. The increase in spreading with Bond number is quite apparent. The increase in spreading with increased radius at $Bo=8.20$ is also shown in figure 9. For $Bo=8.20$, a maximum spreading about four times the gap-width is observed.

As Bond number is found to dominate spreading during mold filling, it is recommended that microgravity experiments be performed to remove gravity effects on the free surface. At $Bo=0$, the effect of Reynolds and Capillary numbers on spreading can be isolated. In particular, the spreading of the fluid along the mold walls due to capillary forces is a phenomena that could be investigated in the absence of gravity. This process is expected to facilitate self impregnation of the mold cavity without an applied injection pressure.

238

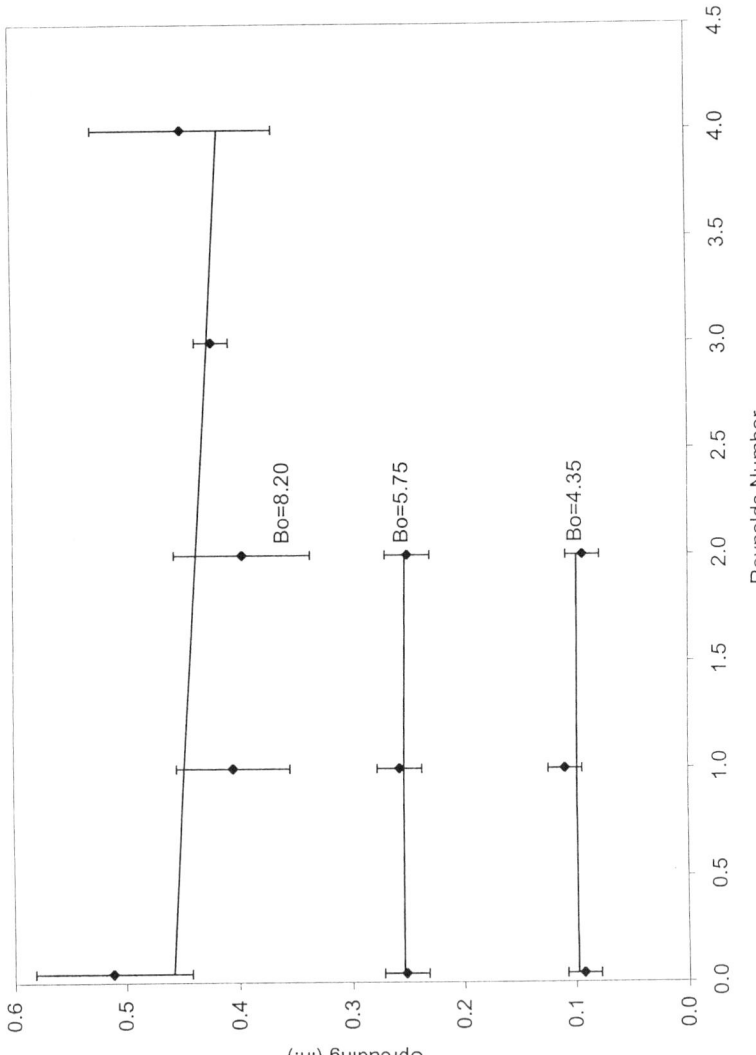

Figure 5: Spreading vs. Reynolds number at R=2.0 in. for three Bond numbers. Ca=0.02.

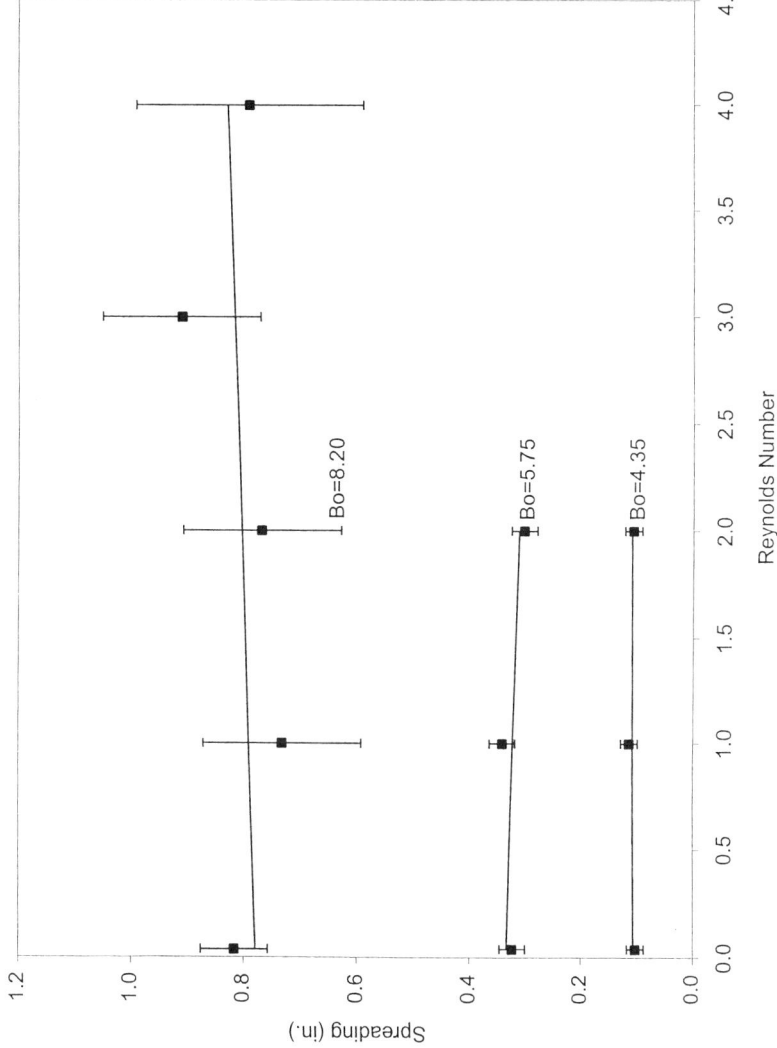

Figure 6: Spreading vs. Reynolds number at R=4.0 in. for three Bond numbers. Ca=0.02.

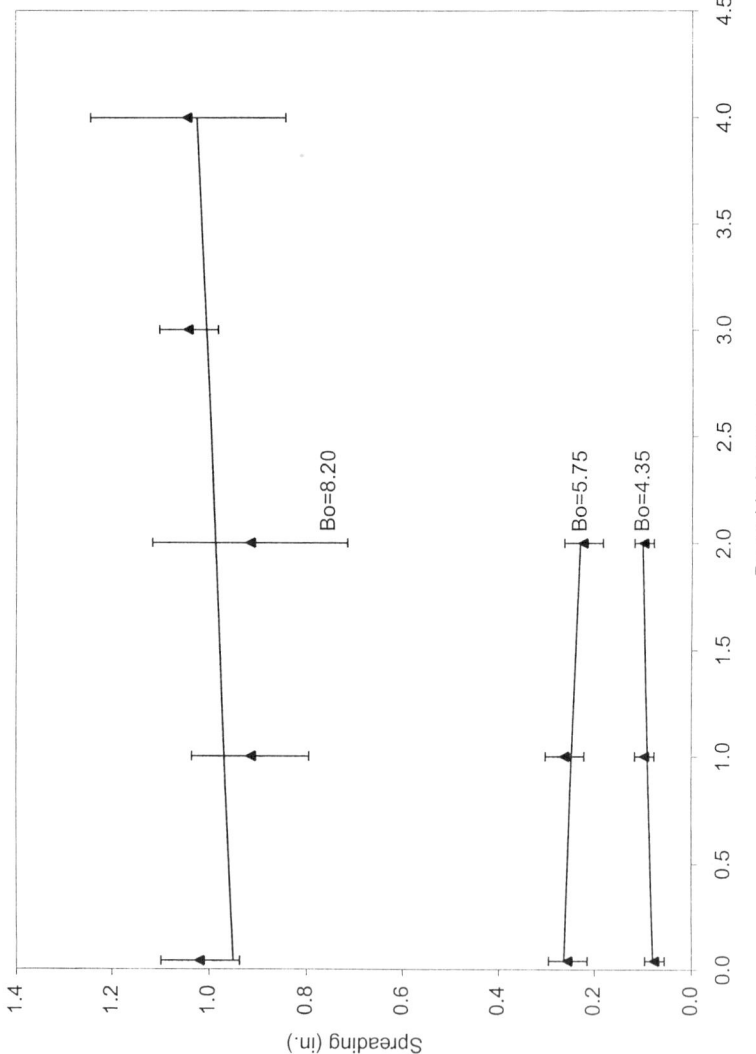

Figure 7: Spreading vs. Reynolds number at R=6.0 in. for three Bond numbers. Ca=0.02.

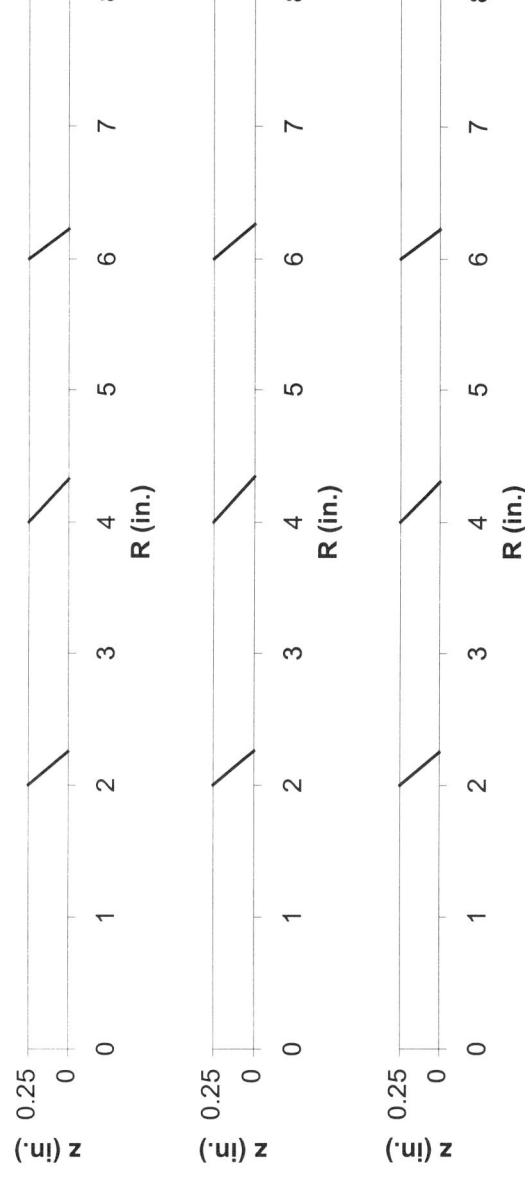

Figure 8: Linear spreading profiles throughout the cavity at Bo=5.75, Ca=0.02. Re=0.04 top, Re=1.0 middle, and Re=2.0 bottom.

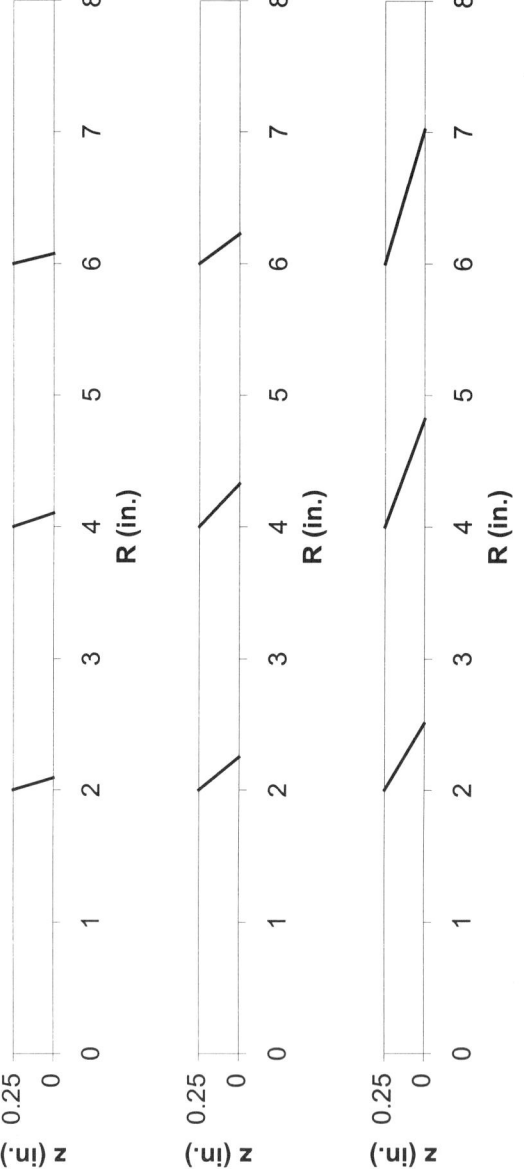

Figure 9: Linear spreading profiles throughout the cavity at Re=0.04, Ca=0.02. Bo=4.35 top, Bo=5.75 middle, and Bo=8.20 bottom.

Additionally, asymmetric fluid front may occur due to capillary spreading along the mold wall containing the inlet which is contacted first by the filling fluid.

Conclusions

Spreading of the fluid front during the filling of a center-gated disk is characterized over a range of fill parameters. The independent process parameters are identified and the effect of individual non-dimensional parameters governing the fluid motion is isolated. Spreading is found to remain nearly constant over the range of Reynolds numbers considered. Bond number is observed to affect the spreading dynamics in a mold filling process drastically. An order in magnitude increase in spreading is measured when the Bond number is increased 100%.

Acknowledgement

This work has been supported by the NASA Microgravity Materials Research Program under Grant NAG8-1271.

References

1. Behrens, R.A., Crochet, M.J., Denson, C.D., and Metzner, A.B. *AIChE J.* **1987**, 33(7), 1178-1186.
2. Jansons, K.M. *J. Fluid Mech.* **1986**, 167, 393-407.
3. Dussan, E.B., and Davis, S.H. *J. Fluid Mech.* **1974**, 65(1), 71-95.
4. Rame, E., and Garoff, S. *J. Colloid Interface Sci.* **1996**, 177, 234-244.
5. Gennes, P.G., Hua, X., and Levinson, P. *J. Fluid Mech.* **1989**, 212, 55-63.
6. Ström, G., Fredriksson, M., Stenius, P., and Radoev, B. *J. Colloid Interface Sci.*, **1989**, 134, 107-116.
7. Blake, J.W. Ph.D. thesis, University of Minnesota, Minneapolis, MN, 1987.
8. Blake, J.W., and Macosko, C.W. *AIChE J.* **1987**, 33(7), 1168-1177.
9. Chen, Q., Rame, E., and Garoff, S. *Phys. Fluids*, **1995**, 7(11), 2631-2639.
10. Hocking, L.M. *J. Fluid Mech.* **1991**, 239, 671-681.
11. Macosko, C.W. *RIM Fundamentals of Reaction Injection Molding*; Hanser: NewYork, NY, 1989; pp 41-42.
12. Olivero, K.A. Masters thesis, University of Oklahoma, Norman, OK, 2000.

Author Index

Ainsworth, William J., 112
Altan, M. C., 227
Apfel, Robert E., 168
Beaudoin, Y., 153
Briskman, V. A., 97
Brown, Kenneth G., 64
Burns, Karen S., 64
Carswell, William E., 51
Chekanov, Yuri A., 112
Crawford, Gregory P., 138
Dao, L. H., 153
Downey, James Patton, 2, 16
El Shall, M. Samy, 185
Emoto, Kazunori, 78
Frazier, Donald O., 51
Gao, Junling, 185
Greenberg, Alan R., 126
Howard, Douglas T., 78
Ingram, Jo, 64
Kennedy, A. P., 217
Kostarev, K. G., 97
Krantz, William B., 126
Masere, Jonathan, 112
Matthews, John T., 78
McLendon, J., 217

McManus, Samuel P., 78
Naumann, Robert J., 51
Nguyen, H. M., 153
Olivero, K. A., 227
Paley, Mark S., 51
Parbhakar, K., 153
Patek, Darayas N., 78
Pekny, Matthew R., 126
Pithawalla, Yezdi B., 185
Pojman, John A., 2, 112
Price, T. Scott, 78
Scholz, Carmen, 203
Sibille, Laurent, 32
Smith, David D., 32
Song, Kwangjin, 168
Sun, Y., 153
Thiruvenkatam, Radhika, 203
Todd, Paul, 126
Upchurch, Billy T., 64
Wessling, Francis C., 78
Whitehead, Joe B., Jr., 138
Wood, George M., 64
Yin, J. M., 153
Zartman, Jeremiah, 126

Subject Index

A

Acceleration
  gravitational, of spacecraft, 18
  gravitational, varying within spacecraft, 22–23
  objects within container experiencing, 18
  platforms for reduction, 18–21
  pressure gradients, 17
  vibrational sources, 23–24
Acceleration reduction
  acceleration sources, 21
  comparison of Shuttle acceleration data with predicted International Space Station (ISS) acceleration environment, 25f
  drop towers, 19
  experimental platforms, 18–21
  limitations of acceleration data, 26
  non-dimensional analysis, 26–28
  parabolic trajectory airplane flights, 20
  quasi-steady and steady acceleration, 22–23
  spacecraft and rockets, 20–21
  transient accelerations, 24, 26
  under-appreciation of platforms for, 29
  vibrational sources of acceleration, 23–24
Acceleration sources
  Active Rack Isolation Sources (ARIS) reducing amplitudes of vibrations, 24
  comparison of Shuttle and predicted International Space Station (ISS) acceleration data, 25f
  limitations of acceleration data, 26
  non-dimensional analysis, 26–28

quasi-steady and steady acceleration, 22–23
  transient, 24, 26
  understanding heat and mass transport in low-g facility, 21
  vibrational sources, 23–24
Acceleration vector, equation, 22
Acrylate monomers
  examples of frontal polymerization, 114
  See also Bubble behavior in frontal polymerization
Active Rack Isolation System (ARIS), reducing amplitudes of vibrations, 24
Airplane flights. *See* Parabolic airplane flights
Aluminum particles
  dependence on particle type, 86, 89
  experiments for aluminized foam, 89, 92
  flaky and granular, 89, 90f
  picture of aluminized polyurethane premix being expelled from mix chamber at 1 atm and near-zero gravity, 91f
  survivability of foams in space, 86
  See also Polyurethane foam production in space
Anomalous diffraction regime, polymer dispersed liquid crystals scattering, 142
Argon bubbles
  following gel formation progress, 102–103
  See also Gel formation under microgravity
Atmospheric drag, source of quasi-steady acceleration, 22
Azotobacter vinelandii UWD. See Bacterial polyesters

B

Bacterial polyesters
  aeration in bioreactor versus shake flask, 211
  analysis methods, 209
  application of poly(β-hydroxybutyrate) (PHB) and poly(β-hydroxybutyrate-co-valerate) (PHBV), 207t
  Azotobacter vinelandii UWD, 205
  bacterial fermentation in bioreactor versus shake flasks, 211–212
  bacteria producing poly(β-hydroxyalkanoates) (PHAs), 206
  biomass production by *A. vinelandii* in flash and bioreactor, 212f
  biopolyesters by bacteria, 206–207
  cell mass accumulating in NASA bioreactor, 208f
  comparison of PHB production by *A. vinelandii UWD* in shake flasks and in NASA bioreactor, 210–214
  drawback of PHA use on earth, 207
  experimental, 208–209
  future work, 215
  glucose consumption, 213–214
  glucose consumption and dissolved oxygen in shake flask and bioreactor, 214f
  growth of *A. vinelandii* grown on minimum medium and polymer yield, 211f
  initial stages of fermentation, 212
  monitoring bacterial activity, 213–214
  oxygen profile, 213–214
  PHB synthesis under microgravity by fermentation of *A. vinelandii UWD* and *Alcaligenes latus*, 209–214
  simulating microgravity using NASA-bioreactor, 207–208
  strain information, 208–209
  structure of PHB, 206f

Bénard convection, heating from below, 4
Benzyl acrylate
  experiments aboard KC-135 aircraft, 115
  temperature profile for frontal polymerization, 114f
  See also Frontal polymerization
Biological systems
  utility in tailoring new materials, 204–205
  See also Bacterial polyesters
Biopolymers, effect of microgravity on porosity and morphology of fibrin and collagen gels, 48
Bond number
  definition, 234–235
  dominating spreading during mold filling, 237
  isolating non-dimensional parameters, 234–236
  spread and flow characteristics during mold filling, 234
  See also Mold filling
Brownian motion, mass transport, 38
Bubble behavior in frontal polymerization
  apparatus, 117, 119
  bubble migration in liquid/liquid system, 121
  bubble nucleation in liquid/liquid system, 120
  diagrams of apparatus used aboard KC-135 aircraft for liquid/liquid experiments, 118f
  experimental liquid/liquid system, 117
  experimental liquid/solid system, 116–117
  experimental viscous liquid/liquid system, 117
  formation of transient periodic bubble structure in liquid/liquid system, 122f

images of foams produced in low and high gravity, 120f
liquid/liquid systems, 120–121
liquid/liquid system under high and low gravity, 121f
liquid/solid system, 119–120
schematics of expected and observed bubble size distribution, 119f
viscous liquid/liquid system, 121–122
viscous liquid/liquid system in high and low gravity, 123f
See also Frontal polymerization
Bulk polymerizations, gravitational effects, 8–9
Buoyancy, influences on foams, 11
Buoyancy-driven convection
Bénard or Rayleigh–Bénard, 4
bulk polymerizations, 8–9
concentration gradient, 4–5
dissolution of polymers, 8
double-diffusive convection, 5
frontal polymerization, 11
heating from below, 4
influence on polymer dispersed liquid crystal morphology, 143
phase separation, 9–10
Rayleigh number, 3–4
Rayleigh–Taylor instability, 5
Rayleigh–Taylor instability with descending front of butyl acrylate polymerization, 11f
schematic of heating single-component fluid, 4f
simultaneous concentration and temperature gradients, 5
single-component fluid, 3–4
Soret effect, 5
tubular reactors, 10
Buoyancy-driven fluid flow, microgravity reducing, 7
Butyl acrylate (BA)
Rayleigh–Taylor instability with descending front of polymerization, 11f, 12
studying poly(BA) fronts, 12

C

Capacitance. See Impedance spectroscopy
Capillary number
definition, 234–235
effect on spreading, 237
isolating non-dimensional parameters, 234–236
spread and flow characteristics during mold filling, 234
See also Mold filling
Cation polymerization, radical. See Gas phase polymerization
Cell structure
closed-cell, 78–79
dynamic decompression and cooling (DDC) foams, 174–175
polyurethane foams, 81, 82f
See also Polyurethane foam production in space
Chemical reactions, gravity, 2
Cluster polymerization
benzene/isoprene and benzene/isoprene/propene clusters, 196
experimental intracluster reactions, 188
experimental set up of cluster source with time-of-flight (TOF) and quadrupole mass spectrometers, 189f
TOF mass spectra benzene/isoprene and benzene/isoprene/propene clusters, 197f
See also Gas phase polymerization
Coagulation kinetics, Smoluchowski theory, 36–37
Coatings, gravitational effects, 9
Collagen gels, effect of microgravity on porosity and morphology, 48
Colloidal crystals, buoyancy-driven convection, 10
Colloidal forces
coupling gravity with, 39
percentage increase in flocculation

rate due to gravity for various particle size ratios, 39, 40f
Commercial refrigerator incubator module (CRIM)
  description, 67–69
  See also Polymerization in microgravity
Composition
  concentration profile for flight sample, 74f
  concentration profile for ground sample, 75f
  flight systems consisting of miscible monomers, 74
  mixing of immiscible monomers during flight, 74–75
Concentration gradients, Soret effect, 5
Conquest I, frontal polymerization experiment, 12
Consort rockets
  experiments demonstrating foam formation, 94
  microgravity effects on polyurethane formation, 83, 86
  microgravity experiment, 89, 92
  photograph of cross-sections of polyurethane samples in microgravity environment versus ground-based comparison, 87f
  photograph of foam beam formed during launch, 95f
  picture of aluminized polyurethane premix being expelled from mix chamber at 1 atm and near-zero gravity, 91f
  See also Polyurethane foam production in space
Containerless processing experiments, *Solidification 1999*, 29
Containers, effect of gravity requiring, for fluids, 29
Continuation equation, rate of gain in mass density, 35
Convection
  Bénard or Rayleigh–Bénard, 4

bulk and surface polymerization with and without, 220
buoyancy-driven, 3–5
concentration gradient, 4–5
diffusion versus, 3
gravity and, 2–3
impedance spectroscopy observing effects of, 222, 224
mass transport in sol-gel system, 35–38
natural, 3
surface-tension driven, 5–7
See also Impedance spectroscopy
Copper particles, laser vaporization-induced polymerization, 196, 198f
Crosslinked polymers, nonlinear optical (NLO) polymers, 154
Crystallization, dynamic decompression and cooling (DDC) foams, 177–178

D

Density
  closed cell foams, 176
  differences between flight- and ground-synthesized samples, 71–72
  foam, as function of process conditions, 175–176
  palladium-containing rods from space mission, 72f
Diacetylenemethylnitroaniline (DAMNA)
  film preparation of poly(DAMNA), 51–52
  structure, 52f, 219f
  ultraviolet polymerization, 52
  See also Impedance spectroscopy; Poly(diacetylene) films
Dielectric spectroscopy
  dielectric constant, 218
  dielectric loss, 218–219
  in situ monitoring polymerization, 218

studying polymerization of various thermoset resins, 218
See also Impedance spectroscopy
Diffusion
convection versus, 3
equation, rate of gain in mass density, 35
mass transport in sol-gel system, 35–38
non-dimensional analysis, 26–27
Dissipation factors. See Impedance spectroscopy
Double-diffusive convection, concentration and temperature gradients, 5
Drag, variation, 22
Drop towers
platforms for acceleration reduction, 19
See also ZARM Bremen drop tower
DVLO (Derjaguin, Landau, Verwey, and Overbeek) model
describing inter-particle surface force potential, 33–34
possible behavior patterns for charged particles in fluid, 34
potential energy, 33
potential energy of attraction, 34
potential energy of repulsion, 35
total potential energy of interaction of sol, 34f
See also Sols
Dynamic decompression and cooling (DDC) process
cell structure of DDC foams, 174–175
crystalline orientation factors for DDC foams, 181f
DDC-induced supercooling, 169–170
description, 169
effect of gravity on foaming, 170–171
scanning electron micrographs of DDC mixed cell foams, open and closed, 179f
supercooling by DDC foaming, 173
tensile properties of DDC closed cell foams, 182f
wide angle X-ray structure pole figures of DDC-produced open and closed foams, 180f
See also Foaming of polyethylenes

E

Electro-optic properties
applications for liquid crystal and polymer dispersions (LCPD), 139
measurement methods, 146
polymer dispersed liquid crystals (PDLC), 142–143
response for polymer dispersed liquid crystal samples fabricated in terrestrial and microgravity environments, 147f, 148f, 149f
schematic of measurement apparatus, 146f
See also Liquid crystal and polymer dispersions (LCPD)
Experimental platforms
drop towers, 19
parabolic trajectory airplane flights, 20
spacecraft and rockets, 20–21
use of Space Shuttle or International Space Station (ISS), 21
See also Acceleration reduction

F

Fibrin gels, effect of microgravity on porosity and morphology, 48
Films
gravitational effects, 9
See also Instrumentation for membrane casting; Poly(diacetylene) films
Flocculation

percentage increase in rate due to gravity for various particle size ratios, 40f
See also Sedimentation
Foaming of polyethylenes
cellular morphology of closed cells, 175
cellular structure of dynamic decomposition and cooling (DDC) foams, 174–175
character of mixed cell morphologies, 179
continuity and momentum balances, 170
crystalline orientation factors, 172
crystalline orientation factors for DDC foams, 181f
crystalline structure and orientation, 180–181
DDC-induced supercooling, 169–170
DDC process, 169
density of closed cell foams, 176
DSC thermograms and crystallinities of as-received resins and the DDC mixed cell foams, 177f
effects of gravity on foaming, 170–171
energy equation, 169–170
foam characterization methods, 172
foam density as function of process conditions, 175–176
growth model, 169–170
influence of reduced gravity environment, 171
materials and foam formation, 171–172
mean densities of low density polyethylene (LDPE) foams as function of process conditions, 176f
mechanical properties, 181–182
microgravity experiment, 182–183
molecular structure and crystallization, 177–178
morphologies of granules and fibers, 178–180

nucleation rate at equilibrium, 170
scanning electron micrographs (SEM) of DDC foams, 174f
SEM micrographs of DDC mixed cell foams, closed and open, 179f
supercooling by DDC foaming, 173
temperature reduction for low-density PE solutions decompressed at 110°C, 173f
tensile properties of DDC closed cell foams, 182f
WAXS method (wide angle X-ray diffraction), 172
WAXS pole figures of DDC-produced open cell poly(butylene terephthalate) (PBT) and closed cell LDPE foams, 180f
Foams
gravitational effects, 11
See also Bubble behavior in frontal polymerization; Polyurethane foam production in space
Free surface shape
common approach for determining, 228
reaction injection molding, 227–228
See also Mold filling
Free volume theory
nonlinear optical (NLO) polymers, 158–162
Simha–Somcynsky free volume, 159
Frontal polymerization
acrylate examples, 114
addition of aqueous blowing agent, 115
advantages and disadvantages of experiments aboard NASA KC-135 aircraft, 115–116
benzyl acrylate experiments aboard KC-135, 115
bubble formation near front and trapping in polymer, 115, 116f
Conquest I rocket flight (1996), 12
defining features, 114
discovery, 11
feasibility of traveling fronts in

solution, 113
gravitational effects, 11–12
localized exothermic reaction zone, 113
Rayleigh–Taylor instability, 11–12
study of bubble interactions on *Conquest I* sounding rocket, 114–115
temperature profile for benzyl acrylate front, 114f
See also Bubble behavior in frontal polymerization

G

Gas phase polymerization
advantages and promises, 186–187
benzene/isoprene gas phase system, 190, 192
benzene/isoprene/propene gas phase system, 193, 196
benzene/propene gas phase system, 192–193
difficulty, 187
early stages of radical cation polymerization of isoprene and propene, 190–196
effect of source pressure on product channels in ternary benzene/isoprene/propene system, 195f
experimental gas phase ion-molecule reactions, 187–188
experimental set-up for Resonant-Two-Photon-Ionization coupled with High Pressure Mass Spectrometry (R2PI–HPMS), 189f
experimental set up for synthesis of nanoparticles coupled with plasma polymerization, 191f
generation of reaction products in benzene/isoprene/Ar system, 192, 194f
implications for, in microgravity, 199–200
initiating method using plasma laser vaporization of metals, 188, 190
initiation, 199
mass spectrum following resonance ionization of benzene in benzene/isoprene/N_2 mixture, 191f
possibilities in gas phase, 186
raw and normalized ion time profiles for reactant and products in benzene/propene system, 194f
scanning electron micrographs (SEM) of polyisobutene containing ultrafine Cu particles, 198f
SEM of polyisobutene microbeads containing Ni nanoparticles, 198f
synthesis of polymeric materials by laser vaporization-induced polymerization, 196, 199
See also Cluster polymerization
Gel, definition, 41
Gel formation under microgravity
cuvettes for experiment, 98–99
defining intensity of convective motion by Grashof parameter, 101
deformation of bubbles, 105, 106f
degassing reaction mixture before filling cuvettes, 101–102
distribution of light intensity behind gas bubble in liquid, 105f
distribution of physico-mechanical properties of gel, 106–107
estimating ability of "Pion-M" to detect gel inhomogeneities, 100
estimating effect of adiabatic conditions on evolution of reaction, 101
experimental equipment, 98–100
experiment "Gel-1" on "Mir" space station, 97
following reaction progress with argon bubbles, 102–103
formation of poly(acrylamide) gel by photoinitiation, 97–98
frontal character, 107
"Gel-1" as launching site for new

series of orbital experiments, 98
general view and schematic of
  cuvette, 99f
interferogram of conversion field
  homogeneities by gas bubbles,
  107f
interferogram of conversion field
  inhomogeneities by formation of
  opaque polymer and gel globules
  in orbital specimen, 107f
number and orientation of
  interference bands, 106, 107f
opaque polymer formation, 102
photographs of gel specimens after
  return to Earth, 103f
photographs of reaction mixture
  during gel formation process in
  microgravity, 104f
"Pion-M" shadow device, 98
polymerization front position versus
  logarithm of time, 105f
reaction mixture, 99–100
reflected light enhancing intensity of
  initiating radiation, 106
shadow pictures of convective
  motion and final structure of gel in
  terrestrial conditions, 101f
solution of analytical equation, 104–
  105
Space Shuttle, 9
spectrum of SD-7 lamp, source of
  initiating radiation, 99, 100f
terrestrial modeling and space
  experiment preparation, 100–102
thermal measurements, 105, 106f
Gelatin, polymerization and
  gelatinization, 8
Gelation
  aggregation behavior of poly(methyl
    methacrylate) (PMMA) spheres,
    47
  behavior of gelling solutions in
    hypergravity, 43
  biopolymers, 48
  composition and molar ratios of
    recipes, 44t

effect of gravity in polymer gels near
  sol-gel transition, 47–48
effect of microgravity on Stöber
  particle growth, 47
gravity effects, 42–48
growth of silica Stöber particles, 46–
  47
microgravity allowing diffusion-
  limited conditions persisting in
  recipes, 44
multiple-gelation behavior, 43
percolation model, 42
phenomenon, 41–42
silica sol-gel structures in 1g and
  microgravity(μg), 45f, 46f
space-grown versus ground-grown
  Stöber particles, 45
Stöber method, 44
studies of silica nanoparticles in
  microgravity on space shuttle
  mission, 43–44
variations of morphology of fibrin
  and collagen gels in microgravity
  flight, 48
Glenn Research Center (GRC),
  NASA's, drop towers, 19
Grashof number
  buoyant force versus viscous force in
    system, 27–28
  defining intensity of convective
    motion, 101
  thermal convection scaling analysis,
    54
Gravity
  affecting relaxation process
    following sudden cooling, 160–
    161
  bubble behavior in liquid/liquid
    system under high and low, 121f
  bulk polymerization, 8–9
  categories for studying polymer
    processing in microgravity, 7
  coatings, films, and membranes, 9
  convection, 2–3
  coupling with colloidal forces, 39
  effect on foaming, 170–171

effects on gelation, 42–48
eliminating pressure head effects, 7
eliminating sedimentation, 7
foams, 11
frontal polymerization, 11–12
influence on chemical reactions, 2
influencing materials processing in solution, 28
objects within container experiencing acceleration, 18
phase separation, 9–10
polymeric foams produced in high and low, 120f
pressure gradients in fluids, 28–29
reducing buoyancy-driven fluid flow, 7
solution effects, 17
tubular reactors, 10
under-appreciation of platforms for acceleration reduction, 29
use of solid containers when working with fluids, 29
viscous liquid/liquid system in high and low, 123f
See also Gelation; Sol-gel systems; Sols
Gravity gradient, steady acceleration experiments in aircraft, 22
Guest-host system
nonlinear optical (NLO) polymers, 154
thermal processing system for sample preparation, 156–157

H

1,6-Hexanedioldiacrylate, frontal polymerization, 116–117
Hexyl acrylate, frontal polymerization, 117
Hydrogen permeability, palladium-containing materials, flight- versus ground-synthesized, 72–73
Hydroxyethylmethacrylate (HEMA), frontal polymerization, 116–117
Hypergravity, behavior of gelling solutions, 43
Hypogravity conditions, parabolic plane flights, 20

I

Immiscible monomers
potential new materials via flight polymerization, 76
See also Polymerization in microgravity
Impedance spectroscopy
bulk and surface photopolymerization of 6-(2-methyl-4-nitroanilino)-2,4-hexadiyn-1-ol (DAMNA), 219
bulk and surface polymerization with and without convection, 220f
capacitance for bulk solution polymerization of polydiacetylene in various orientations, 224f
capacitance for surface and bulk solution at 1 kHz, 221f
capacitance for surface polymerization of polydiacetylene in horizontal orientation, 223f
capacitance versus frequency at various times, 225f
change in capacitance versus frequency, 224
comparing surface and bulk polymerizations, 221
dissipation factor for bulk solution polymerization of polydiacetylene in various orientations, 225f
dissipation factor for surface polymerization of polydiacetylene in horizontal orientation, 223f
dissipation factor surface and bulk solution polymerization at 1 kHz, 222f
experimental, 220

in situ monitoring technique, 218
observing convective effects, 222, 224
results for solution polymerization at various orientations, 222
structure of DAMNA monomer, 219f
studying polymerization of variety of thermoset resins, 219
temperature controlled cell, 220
Initiation
cluster polymerization, 188
gas phase polymerization, 187–188
plasma laser vaporization of metals, 188, 190
Instrumentation for membrane casting
diagram of Membrane Casting Apparatus (MCA-1) for application in microgravity, 132f
dry-casting experiments, 129–130
elimination of "sloshing" problem, 132
establishing conditions for casting polymeric membranes on orbital space flights, 130
initial sliding-block mechanism for casting polymer and ceramic membranes in low gravity, 129f
light reflectance versus time recorded using optical reflectometry system, 135f
Materials Dispersion Apparatus (MDA), 129–130
MCA-1 operation, 132–133
overall design of MCA-1, 131–132
requirements for extended MCA designs, 135
rocket-borne experiments, 129
self-contained optical sensor-recorder for MCA-1, 133–134
sketch of assembled construction of casting station of MCA-1, 131f
sketch of optical reflectometry system, 134f
sliding-block casting apparatus, 130–131

sloshing of polymer solution before phase inversion, 133f
International Space Station (ISS)
comparison of Shuttle acceleration data with predicted ISS acceleration environment, 25f
exploration in future, 204
minimizing effect of vibrations on experiments, 24
platforms for acceleration reduction, 20–21
Isoprene
intracluster reactions for benzene/isoprene and benzene/isoprene/propene, 196, 197f
radical cation polymerization of benzene/isoprene gas phase system, 190, 192
radical cation polymerization of benzene/isoprene/propene gas phase system, 193, 196
See also Gas phase polymerization

K

Kamisunagawa, Japan, drop tower, 19
KC-135 parabolic flights. *See* Bubble behavior in frontal polymerization
Kinetics
microgravity thin film growth rate equation, 56–58
Smoluchowski theory of coagulation, 36–37

L

Laser vaporization, inducing polymerization, 196, 199
Liquid crystal and polymer dispersions (LCPD)
characterization methods of microgravity and terrestrial

samples, 146
chemical structure for nematic monomer RM257, 145f
composition and structures of monomers (K15, K21, M24, and T15) in nematic liquid crystal mixture (E7), 144f
droplet morphology, 147, 149
electro-optic applications, 139
electro-optic properties, 142–143
electro-optic response for E7/NOA65 sample fabricated in microgravity environment, 148f
electro-optic response for E7/NOA65 sample fabricated in terrestrial environment, 147f
electro-optic responses for terrestrial and microgravity E7/NOA65 PDLCs, 149
experimental, 144–146
high resolution scanning electron images of surfaces of terrestrial and microgravity polymer networks, 150f
holographically formed polymer dispersed liquid crystals (H-PDLC), 139, 140f
influence of buoyancy driven convection on PDLC morphology, 143
investigating gravitational influence on morphology, 143
microgravity, 143–144
mixture preparation, 144–145
Monte Carlo simulations for investigating effect of polymerization-induced phase separation (PIPS) on PDLCs, 143
morphological influence on electro-optic properties, 147, 148f
morphology range, 139
optical micrographs of polymer network in E7/RM257 NSLC fabricated in terrestrial and microgravity environments, 150f
optical microscopy and scanning electron microscopy of NSLC samples processed in microgravity, 149–150
phase-separated morphology of network stabilized liquid crystals (NSLC), 140, 141f
photo-polymerization of PDLC and NSLC mixtures, 145–146
polymerization-induced phase separation (PIPS), 141–142
scattering in Rayleigh–Gans or Anomalous Diffraction regime, 142
schematic of electro-optic measurement apparatus, 146f
schematic of H-PDLC sandwiched between transparent conducting electrodes, 140f
schematic of macroscopic alignment of RM257 molecules by K15 molecules due to rubbed polyimide, 145f
schematic of NSLC sandwiched between transparent conducting electrodes, 141f
schematic of PDLC sandwiched between transparent conducting electrodes, 140f
solvent-induced phase separation (SIPS), 141
switching time, 142
switching voltage, 142
thermally induced phase separation (TIPS), 141
turn-off time, 142
turn-on time, 142
uses, 139
Liquids, deformations with applied force, 17–18
Low density polyethylene (LDPE). *See* Foaming of polyethylenes
Low gravity. *See* Instrumentation for membrane casting

M

Macrovoids
  growth using low-gravity methods, 127–128
  See also Instrumentation for membrane casting
Marangoni convection
  buoyancy–driven convection and, in system with free interface, 6–7
  phase separation, 9–10
Marangoni effects, studying processes, 7
Marangoni number
  definition, 6
  first power of layer depth, 7
  surface tension analog of Grashof number, 28
  variation in interfacial tension, 6
Membrane industry
  markets, 126–127
  worldwide competitiveness, 127
Membranes
  gravitational effects, 9
  macrovoid (MV) pores, 127
  phase separation/inversion process, 127
  prior studies of casting in low gravity, 128–129
  technologies in industry, 126–127
  See also Instrumentation for membrane casting
Metal particles, laser vaporization-induced polymerization, 196, 199
Methyl methacrylate (MMA), photopolymerization during parabolic airplane flights, 8
Microcrystalline arrays, potential for microgravity affecting polymers, 65
Microgravity
  diffusion-limited conditions persisting in sol-gel solutions, 44
  diverse research, 29
  gelation studies of silica nanoparticles during space shuttle mission, 43–44
  mechanisms affecting polymerization, 218
  studying polymer processing in, 7
  terminology, 17
  See also Acceleration reduction; Gel formation under microgravity; Gravity; Impedance spectroscopy; Liquid crystal and polymer dispersions (LCPD); Poly(diacetylene) films; Polymerization in microgravity; Polyurethane foam production in space; Sol-gel systems
Microorganisms, utility in tailoring new materials, 205
Mir space station. *See* Gel formation under microgravity
Mold filling
  Bond and Reynolds numbers for eleven sets of experiments, 236*t*
  Bond number dominating spreading during filling, 237
  common approach in determining free surface shapes, 228
  defining non-dimensional parameters at location of first fluid front probe, 235
  effect of Reynolds and Capillary numbers on spreading, 237
  equation for spreading, 234
  experimental mold cavity diagram, 231*f*
  experimental set-up, 230, 234
  experiments (eleven) at various Bond and Reynolds numbers, 236
  isolating non-dimensional parameters, 234–236
  linear spreading profiles throughout cavity at three Bond numbers with Reynolds and Capillary numbers constant, 242*f*
  linear spreading profiles throughout cavity at three Reynolds numbers with Bond and Capillary numbers

constant, 241*f*
reaction injection molding (RIM), 227–228
recommending microgravity experiments for removing gravity effects on free surface, 237, 243
representative fluid front shapes, 229*f*
Reynolds, Bond, and Capillary numbers, 234–236
Reynolds number governing flow, 228
schematic cross section of probe wiring depicting advancement of fluid front through probe location, 233*f*
set-up for mold filling experiments, 232*f*
spreading of bottom of fluid front due to gravity, 229*f*
spreading results for eleven sets of experiments, 237
spreading versus Reynolds number (R=2.0) for three Bond numbers, 238*f*
spreading versus Reynolds number (R=4.0) for three Bond numbers, 239*f*
spreading versus Reynolds number (R=6.0) for three Bond numbers, 240*f*
symmetric fluid front in absence of gravity effects, 229*f*
trend in using lower viscosity resins, 228
Molecular order, potential for microgravity affecting, of polymers, 65
Molecular structure, dynamic decompression and cooling (DDC) foams, 177–178
Monte Carlo simulations, polymer dispersed liquid crystals, 143–144
Morphology
effect of microgravity on fibrin and collagen gels, 48
granules and fiber morphologies of foams, 178–180
See also Liquid crystal and polymer dispersions (LCPD)
Multiple-gelation behavior, gelation interface, 43

N

NASA bioreactor. *See* Bacterial polyesters
Navier–Stokes equation
conservation of momentum in sol-gel system, 35
neglecting inertia terms to get Stokes equation, 38
Network stabilized liquid crystals (NSLC). *See* Liquid crystal and polymer dispersions (LCPD)
Nickel nanoparticles, laser vaporization-induced polymerization, 198*f*, 199
Non-dimensional analysis
diffusion, 26–27
example of fluid in rectangular enclosure with thermal gradient, 27–28
Grashof number, 27–28
Marangoni number, 28
Strouhal number, 28
Nonlinear optical (NLO) polymers
ability to tailor materials, 155
ceramic cell support, 156
characterization (relaxation of second harmonic signal), 157
drop capsule for ZARM Bremen drop tower, 155–157
effect of cross-linking reaction on relaxation process, 162
enhancement of second harmonic (SH) signal for sample preparation under microgravity, 164, 165*t*
experimental, 155–157
experiments at ZARM drop tower facility, 163

free volume and simulations, 158–162
gravity accelerating structure relaxation process for sudden cooling, 161
gravity affecting relaxation process following sudden cooling, 160–161
key problems, 154–155
kinetics of structure relaxation process following sudden cooling or heating, 159–161
layout of specimens and typical parameters in thermal NLO processing of Disperse Red 1/poly(methyl methacrylate) (DR1/PMMA) films, 164t
molecular design approaches, 154
NLO devices fabrication, 157
photo crosslinked NLO processing of films, 163–164
Photo Processing System (platform 4), 156
poled-polymer systems, 154
relaxations, 155
sample holder with NLO devices, 156f
second harmonic generation (SHG) intensity for samples processes at 10^{-6} g and 1 g, 164, 165t
Simha–Somcynsky equation of state, 158
Simha–Somcynsky free volume, 159
sudden cooling, 161, 162f
Thermal Processing System for sample preparation in Guest Host system, 156–157
Vogel–Tamann–Fulcher (VTF) equation, 155
volume recovery process of poly(vinyl acetate) after sudden cooling for different cross-linking rate parameter, 163f
volume recovery process of poly(vinyl acetate) after sudden cooling for different gravity parameter, 162f
volume relaxation of poly(vinyl acetate) after sudden cooling, 161, 162f

O

Orbital experiments. *See* Gel formation under microgravity
Order, molecular, potential for microgravity affecting polymers, 65

P

Parabolic airplane flights
photopolymerization of methyl methacrylate, 8
platforms for acceleration reduction, 20
See also Bubble behavior in frontal polymerization
Percolation model, gelation phenomenon, 42
Permeability, palladium-containing flight-synthesized samples, 72–73
Phase separation, gravitational effects, 9–10
Pion-M
shadow device, 98, 100
See also Gel formation under microgravity
Plasma polymerization
experimental, 188, 190
initiation by laser vaporization, 196, 199
See also Gas phase polymerization
Poled-polymer systems, 154
Poly(acrylamide) (PAA) gels
formation by photoinitiation, 97–98
microgravity gelation experiment, 42–43
properties for flight- and ground-based, 65
See also Gel formation under

microgravity
Poly(butylene terephthalate) (PBT)
  crystalline orientation factors for
    PBT foam, 181f
  crystalline structure, 180
  open cell structures, 179
  See also Foaming of polyethylenes
Poly(diacetylene)
  effects of convection during
    photodeposition, 8
  See also Impedance spectroscopy
Poly(diacetylene) films
  achieving diffusion-controlled
    transport conditions, 53
  aggregation process, 55
  assumption of pseudo-zeroth order
    conditions making scaled
    experiment unacceptable, 56
  change in concentration due to
    diffusive feeding of depleting
    interface, 56–57
  convection reduction for diffusive
    mass transport domination in
    system, 54
  diacetylenemethylnitroaniline
    (DAMNA) monomer, 52f
  diffusive/kinetic rate equation for
    thin film growth applicable to
    microgravity, 58
  discarding classical thermally stable
    top-heated configuration, 53
  discrepancy between thickness
    grown in space and predictions,
    60, 62
  experimental conditions for ground-
    control experiments, 59t
  films of poly(DAMNA), 51–52
  film thickness as function of time in
    microgravity, 58f
  Grashof number for thermal
    convection scaling analysis, 54
  impossible to use scaling approach
    for convection-free experiments in
    laboratory, 56
  initial stages of growth, 54
  interfacial concentration, 57
  kinetic rate equation for thin film
    growth, 56
  microgravity justification, 53–56
  microgravity polymer thin film
    growth (PTFG) hardware, 61f
  microgravity thin film growth
    experiment, 60
  microgravity thin film growth rate
    equation, 56–58
  polymerization in solution reducing
    film growth rate, 62
  rate of change of thickness or growth
    rate of film, 57
  results from microgravity flight
    experiments, 62t
  results from microgravity ground-
    control experiments, 62t
  thin film growth schematic, 52f
  ultraviolet (UV) polymerization
    process for film growth, 52
  UV intensity as function of
    penetration depth in PTFG cell,
    55f
Polyesters. See Bacterial polyesters
Poly(ethylene)
  1,2-dichlorotetrafluoroethane as
    foaming gas, 168
  materials and formation, 171–172
  See also Foaming of polyethylenes
Poly(β-hydroxybutyrate) (PHB). See
  Bacterial polyesters
Polymer dispersed liquid crystals
  (PDLC)
  buoyancy effects, 10
  See also Liquid crystal and polymer
    dispersions (LCPD)
Polymeric foams
  commercial production, 112
  first polyurethane foam in reduced
    gravity, 113
  foaming mechanism, 168
  gases as blowing agent, 168
  uses, 112
  See also Bubble behavior in frontal
    polymerization; Foaming of
    polyethylenes; Polyurethane foam

production in space
Polymerization
  early stages in gas phase, 186
  limited fundamental understanding in solution, 185–186
  microgravity affecting mechanisms, 218
  See also Gas phase polymerization
Polymerization-induced phase separation (PIPS)
  fabrication of liquid crystal and polymer dispersions, 141–142
  See also Liquid crystal and polymer dispersions (LCPD)
Polymerization in microgravity
  analysis methods, 67
  assembled test tube, post-launch, 67*f*
  composition, 74–75
  concentration profile for flight sample, 74*f*
  concentration profile of ground-synthesized materials, 75*f*
  construction of polymerization module (PM), 68
  cut-out view of commercial refrigerator incubator module (CRIM) showing PM loaded with test tubes, 69*f*
  cylindrical rods as sample shape, 66
  density, 71–72
  density of palladium containing rods from second mission, 72*f*
  difference between flight and ground samples, 76
  hydrogen permeability, 72–73
  hydrogen permeability for palladium containing materials, 73*t*
  immiscible systems, 70–71
  instrumentation, 67–69
  materials, 66–67
  miscible systems, 70
  mixing of immiscible monomers, 74–75
  near-zero gravity environment, 65
  oxygen permeability, 73
  permeability, 72–73
  poly(acrylamide) gels exhibiting properties different from ground-based, 65
  poly(ethylene glycol) 400 methacrylate (PEG-400) and disiloxane, 75
  polymers from immiscible monomers, 76
  potential for affecting molecular order, 65
  preprogrammed temperature profile aboard CRIM, 70
  procedure, 69–70
  Spacehab® laboratory, 65–66
  temperature profile for typical mission, 71*f*
Polymerization module (PM), description, 67–69
Polymer processing in microgravity
  bulk polymerization, 8–9
  buoyancy-driven and Marangoni convection in system with free interface, 6–7
  buoyancy-driven convection, 3–5
  categories for studying, 7
  chemical concentration and temperature affecting surface tension, 6*f*
  coatings, films, and membranes, 9
  convection, 2–3
  convection by concentration gradient, 4–5
  double-diffusive convection, 5
  eliminating pressure head effects, 7
  eliminating sedimentation, 7
  foams, 11
  frontal polymerization, 11–12
  gravity and convection, 2–3
  heating from below, 4
  Marangoni number, 6
  phase separation, 9–10
  Rayleigh number, 3–4
  Rayleigh–Taylor instability with descending front of butyl acrylate polymerization, 11*f*

reducing buoyancy-driven fluid flow, 7
schematic of heating single-component fluid, 4f
simultaneous concentration and temperature gradients, 5
surface-tension driven convection, 5–7
tubular reactors, 10
Poly(methyl methacrylate) (PMMA)
aggregation behavior of PMMA spheres, 47
See also Nonlinear optical (NLO) polymers
Polysulfone membranes, effect of acceleration level, 9
Polyurethane foam production in space
cell structure, 81, 82f
closed-cell foams, 78–79
Consort-series research rockets, 83, 86
Consort-series rockets demonstrating foam formation, 94
dependence on aluminum particle type, 86, 89
digitized representations, 88f
effects of aluminum particles, 86, 89
effects of vacuum of space, 92, 94
gravitational effects on foam formation and structure, 81, 83
high gravity effects, 81, 83
microgravity effects, 83, 86
microgravity experiment, 79
microgravity experiment to form aluminized foam, 89, 92
models of Kelvin and Weaire–Phelan cell structure, 79, 82f
photograph of cross-section, 82f
photograph of cross-sections of laboratory samples with granular and flaky aluminum, 90f
photograph of cross-sections of samples from microgravity environment in sounding rocket experiment versus ground-based, 87f
photograph of foam beam formed during Consort launch, 95f
photographs at X10 magnification of foams under various laboratory conditions, 84f, 85f
photographs of foam beams prepared with different polymer mixes in vacuum in laboratory, 93f
photomicrograph (100X) of sample prepared with flaky aluminum, 90f
picture of aluminized premix being expelled from mix chamber at 1 atm and near-zero gravity during Consort rocket flight, 91f
space as harsh environment, 80
space as reaction environment, 79–80
Poly(vinyl acetate), volume relaxation after sudden cooling, 161, 162f, 163f
Porosity, effect of microgravity on fibrin and collagen gels, 48
Potential energy
attraction, 34
DLVO (Derjaguin, Landau, Verwey, and Overbeek) model, 33
repulsion, 35
repulsive double-layer force, 33
total, of interaction of sol, 34f
See also Sols
Pressure, gradients in fluids, 28–29
Pressure head effects, microgravity eliminating, 7
Propene
intracluster reactions for benzene/isoprene/propene, 196, 197f
radical cation polymerization of benzene/isoprene/propene gas phase system, 193, 196
radical cation polymerization of benzene/propene gas phase system, 192–193
See also Gas phase polymerization

R

Radical cation polymerization. *See* Gas phase polymerization
Rayleigh–Bénard convection, heating from below, 4
Rayleigh–Gans regime, polymer dispersed liquid crystals scattering, 142
Rayleigh number
　bulk convection, 7
　dimensionless quantity, 3–4
Rayleigh–Taylor instability
　descending front of butyl acrylate polymerization, 11f
　descending fronts, 11–12
　more dense fluid on top of less dense fluid, 5
Reaction injection molding (RIM)
　curing and mold filling, 227–228
　trend for use of low viscosity resins, 228
　See also Mold filling
Resonant-Two-Photon-Ionization coupled with high pressure mass spectrometry (R2PI–HPMS)
　experimental set-up, 189f
　gas phase polymerization, 187–188
Reynolds number
　definition, 234–235
　effect on spreading, 237
　governing flow, 228
　isolating non-dimensional parameters, 234–236
　spread and flow characteristics during mold filling, 234
　See also Mold filling
Rockets, platforms for acceleration reduction, 20–21

S

Sedimentation
　gravitational, 39–40
　mass transport, 38
　microgravity eliminating, 7
　percentage increase in flocculation rate due to gravity for various particle size ratios, 40f
Shake flasks. *See* Bacterial polyesters
Side-chain, nonlinear optical (NLO) polymers, 154
Smoluchowski theory
　coagulation kinetics, 36
　coagulation time, 37
　irreversible coagulation, 37f
Sol-gel systems
　effect of gravity in polymer gels near transition, 47–48
　growth of silica Stöber particles, 46–47
　mass transport: convection and diffusion, 35–38
　silica sol-gel structures at 1g and microgravity (μg), 45f, 46f
　space-grown versus ground-grown Stöber particles, 45
　See also Gelation; Sols
Sol-gel transition, classical theory and percolation model, 41–42
Solidification 1999, containerless processing experiments, 29
Solids, resisting deformation by shearing force, 17
Sols
　change in singlet population, 36–37
　coagulation time, 37
　continuation equation, 35
　coupling of gravity with colloidal forces, 39
　diffusion equation, 35
　DVLO (Derjaguin, Landau, Verwey, and Overbeek) model describing inter-particle surface force potential, 33–34
　effects of convective flows with respect to diffusive mass transport, 36
　gravitational stability, 39–40
　mass transport: convection and diffusion, 35–38

mass transport: sedimentation and Brownian motion, 38
Navier–Stokes equation, 35
percentage increase in flocculation rate due to gravity for various size ratios, 40f
population of higher order aggregates, 37f, 38
possible behavior patterns, 34
potential energy of attraction V_A, 34
potential energy of repulsion V_S, 35
potential energy V_R, 33
Stokes equation, 38
total potential energy of interaction, 34f
velocity of fluid surrounding particle, 38
Von Smoluchowski coagulation kinetic theory, 36
Von Smoluchowski theory of irreversible coagulation, 37f
Solvent-induced phase separation (SIPS), fabrication of liquid crystal and polymer dispersions, 141
Soret effect, concentration gradients, 5
Space
  effects of vacuum of, 92, 94
  harsh environment, 80
  photographs of foam beams using different polymer mixes in vacuum in laboratory, 93f
  reaction environment, 79–80
  See also Polyurethane foam production in space
Spacecraft
  gravitational acceleration, 18
  gravitational acceleration varying within, 22–23
  platforms for acceleration reduction, 20–21
  transient accelerations, 24, 26
Spacehab®
  laboratory development, 65–66
  See also Polymerization in microgravity
Space Shuttle
  comparison of Shuttle acceleration data with predicted International Space Station (ISS) acceleration, 25f
  exploration in future, 204
  gelation studies of silica nanoparticles in microgravity, 43–44
  large-particle size monodisperse latexes, 10
  NASA bioreactor for simulating microgravity, 207–208
  platforms for acceleration reduction, 20–21
  polymer gel formation, 9
  polysulfone membrane formation, 9
  transient accelerations, 24, 26
  See also Polymerization in microgravity
Spreading
  equation, 234
  significance for different fill parameters, 237
  See also Mold filling
Stability, gravitational, of sols, 39–40
Stöber particles
  effect of microgravity on growth, 47
  growth of silica, 46–47
  method for particle growth, 44
  space-grown versus ground-grown, 45
  See also Gelation; Sol-gel systems
Stokes equation, mass transport, 38
Strouhal number, evaluating effects of vibrations, 28
Styrene and 4-vinyl pyridine, flight- versus ground-synthesized polymers, 74–75
Supercooling
  dynamic decompression and cooling (DDC) process, 173
  See also Foaming of polyethylenes
Surface-tension driven convection
  chemical concentration and temperature affecting surface tension, 6f

gradients in interfacial tension at interface of fluids, 5–7
Marangoni number, 6

T

Tetraorthosilicate (TEOS)
  composition and molar ratios of sol-gel recipes, 44t
  gels in microgravity at low TEOS concentration, 47
  preparation, 44
  See also Sol-gel systems
Thermally induced phase separation (TIPS), liquid crystal and polymer dispersions, 141
Thin films
  microgravity growth experiment, 60
  microgravity thin film growth rate equation, 56–58
  thickness as function of time in microgravity, 58f
  See also Poly(diacetylene) films
Tubular reactors, gravitational effects, 10

V

Vacuum
  effect of, of space, 92, 94
  photographs of polyurethane foam beams prepared in vacuum in laboratory, 93f
Vibrational sources
  acceleration, 23–24
4-Vinyl pyridine and styrene, flight-versus ground-synthesized polymers, 74–75
Vogel–Tamann–Fulcher (VTF) equation, relating viscoelastic properties of polymers to glass transition, 155
Von Smoluchowski, coagulation kinetics, 36–37

Z

ZARM Bremen drop tower
  experimental, 155–157
  fabricating nonlinear optical (NLO) polymers, 163–165
  See also Nonlinear optical (NLO) polymers
Zero Gravity Research Facility, drop tower, 19